WATER RESOURCE PLANNING, DEVELOPMENT AND MANAGEMENT

NATIONAL WATER SUSTAINABILITY

THE WATERSMART PROGRAM

WATER RESOURCE PLANNING, DEVELOPMENT AND MANAGEMENT

Additional books in this series can be found on Nova's website under the Series tab.

Additional E-books in this series can be found on Nova's website under the E-book tab.

WATER RESOURCE PLANNING, DEVELOPMENT AND MANAGEMENT

NATIONAL WATER SUSTAINABILITY

THE WATERSMART PROGRAM

WILLLIAM OWENS
EDITOR

New York

For permission to use material from this book please contact us:
Telephone 631-231-7269; Fax 631-231-8175
Web Site: http://www.novapublishers.com

Additional color graphics may be available in the e-book version of this book.

Library of Congress Cataloging-in-Publication Data

ISBN: 978-1-62618-103-8

Published by Nova Science Publishers, Inc. † New York

CONTENTS

PREFACE

Adequate supplies of water are essential to people, the economy and the environment. Yet the Nation faces an increasing set of water resource challenges. Aging infrastructure, rapid population growth, depletion of groundwater resources, impaired water quality associated with particular land uses and covers, reservoir sedimentation, water needed for human and environmental uses, increased domestic energy development, and climate variability and change all play a role in determining the amount of fresh water available at any given place and time. It is increasingly recognized that water is the primary means through which climate change impacts the earth and people's livelihoods and well being. This book examines the WaterSMART initiative, a program that recognizes the U.S. Department of Interior's role in providing Federal leadership in the pursuit of sustainable water supply for people, the economy, and the environment.

Chapter 1 – Collaborative partnerships that go beyond political and institutional jurisdictions must be developed to ensure that the Nation's limited water resources are used efficiently, sufficient amounts are retained to protect and restore the environment, and supplies are managed to reliably meet new demands. Future generations depend upon the leadership we provide today to ensure that they have opportunities to prosper and to enjoy and benefit from one of our Nation's most important resources -- fresh, clean water. The Federal government must provide leadership and assistance, but cannot do it alone. States, tribes, local governments, private associations, and organizations representing the citizens and resource users also must join in leading towards a sustainable water resources future.

Past divisions must be broken down and new relationships forged with policies and programs that recognize the many needs, resource limitations and opportunities, and uncertainty.

To ensure that the Department of the Interior is positioned to meet these challenges, the Secretary issued an order (Secretarial Order 3297) in February 2010 establishing the WaterSMART (Sustain and Manage America's Resources for Tomorrow) Program.

WaterSMART will work with states, tribes, local governments, and non-governmental organizations to secure and stretch water supplies for use by existing and future generations to benefit people, the economy, and the environment, and will identify adaptive measures needed to address climate change and future demands.

Section 5(a) of the Secretarial Order calls for the development of a written strategic plan to implement the WaterSMART Program.

This DOI WaterSMART Strategic Implementation Plan fulfills that requirement and will provide the framework the Department of the Interior will use to provide Federal leadership in moving toward a sustainable water resources future.

Chapter 2 - The health, security, and economic well-being of the American people depend on adequate supplies of clean water. We use water to drink and stay healthy, produce energy, grow food, advance the economy, and engage in recreation. On a more basic level, water is vital to maintaining the healthy ecosystems that support our lives. But the amount of water available to support people, the economy, and the environment is finite. As our need for water continues to grow, available supplies will be increasingly strained, risking the depletion of our shared natural resources and raising the potential for conflict.

Three years ago, Congress passed the SECURE Water Act of 2009, directing the United States Department of the Interior to develop a sustainable water management policy. In 2010, Secretary of the Interior Ken Salazar established WaterSMART, combining existing programs with new initiatives to create a broad framework for wisely managing the Nation's water supplies. Through these programs, Interior is actively working with Tribal, State, regional, and local water managers to address a range of urgent issues associated with water scarcity.

In: National Water Sustainability
Editor: Willliam Owens
ISBN: 978-1-62618-103-8

Chapter 1

DOI WATERSMART STRATEGIC IMPLEMENTATION PLAN*

U.S. Department of the Interior

SECTION 1. A SUSTAINABLE WATER RESOURCES FUTURE

Adequate supplies of water are essential to people, the economy and the environment. Yet the Nation faces an increasing set of water resource challenges.

Aging infrastructure, rapid population growth, depletion of groundwater resources, impaired water quality associated with particular land uses and covers, reservoir sedimentation, water needed for human and environmental uses, increased domestic energy development, and climate variability and change all play a role in determining the amount of fresh water available at any given place and time. It is increasingly recognized that water is the primary means through which climate change impacts the earth and people's livelihoods and well being. Water shortage and water-use conflicts have become more commonplace in many areas of the United States.

Collaborative partnerships that go beyond political and institutional jurisdictions must be developed to ensure that the Nation's limited water

* This is an edited, reformatted and augmented version of the U.S. Department of the Interior publication, dated March 22, 2011.

* This is an edited, reformatted and augmented version of the U.S. Department of the Interior publication, dated March 22, 2011.

resources are used efficiently, sufficient amounts are retained to protect and restore the environment, and supplies are managed to reliably meet new demands. Future generations depend upon the leadership we provide today to ensure that they have opportunities to prosper and to enjoy and benefit from one of our Nation's most important resources -- fresh, clean water. The Federal government must provide leadership and assistance, but cannot do it alone. States, tribes, local governments, private associations, and organizations representing the citizens and resource users also must join in leading towards a sustainable water resources future.

Past divisions must be broken down and new relationships forged with policies and programs that recognize the many needs, resource limitations and opportunities, and uncertainty.

To ensure that the Department of the Interior is positioned to meet these challenges, the Secretary issued an order (Secretarial Order 3297) in February 2010 establishing the WaterSMART (Sustain and Manage America's Resources for Tomorrow) Program.

WaterSMART will work with states, tribes, local governments, and non-governmental organizations to secure and stretch water supplies for use by existing and future generations to benefit people, the economy, and the environment, and will identify adaptive measures needed to address climate change and future demands.

Section 5(a) of the Secretarial Order calls for the development of a written strategic plan to implement the WaterSMART Program.

This DOI WaterSMART Strategic Implementation Plan fulfills that requirement and will provide the framework the Department of the Interior will use to provide Federal leadership in moving toward a sustainable water resources future.

SECTION 2. PROGRAM COORDINATION

Section 4(a) of Secretarial Order 3297 provides that the Department of the Interior (DOI) will coordinate across agencies and programs to implement water sustainability efforts. Secretarial Order 3297 also lists specific programs, including the Bureau of Reclamation's Title XVI, Basin Study Program, and WaterSMART Grants, and the U.S. Geological Survey's (USGS) WaterSMART Availability and Use Assessments. All WaterSMART programs recognize that, while the Federal government can provide leadership, working with states, tribes, local governments, and non-governmental organizations (NGO) will lead to greater accomplishments towards a sustainable future. Additionally, within the Federal government DOI will coordinate activities with the U.S. Department of Agriculture (USDA), U.S. Department of Commerce, U.S. Department of Energy (DOE), and the Environmental Protection Agency (EPA). Coordination with states will leverage existing entities such as the Western States Federal Agency Support Team (WestFAST).

All DOI bureaus are working with partners within the Landscape Conservation Cooperatives (LCCs) and Climate Science Centers (CSCs), identified within Secretarial Order 3289, which include outreach and coordination activities with water, fish, wildlife, cultural, and land resources. This section of the Secretarial Order will be implemented through the following activities.

a. *Assistant Secretary for Policy, Management, and Budget*
   I. The Assistant Secretary for Policy, Management and Budget is DOI's Senior Sustainability Officer and chairs DOI's Sustainability Council with the responsibility, among others, to improve water use efficiency and management at DOI's owned and managed facilities. The Sustainability Council includes cross-functional representation from DOI senior managers, bureau and office program managers, and technical work groups for sustainable practices and environmental compliance.
b. *Assistant Secretary for Water and Science*
   I. The office of the Assistant Secretary for Water and Science is responsible for ensuring implementation of all aspects of Secretarial Order 3297.
   II. The office of the Assistant Secretary of Water and Science is leading the implementation of the Cooperative Watershed Management Program, authorized by the Omnibus Public Land Management Act of 2009, P.L. 111-11 (P.L. 111-11), which brings together local watershed stakeholders to design and implement projects that enhance water conservation, improve water quality, increase ecological resiliency, and reduce the potential for water conflict. This program will include DOI bureaus as well as state and local governments and other interested entities.
c. *Fish and Wildlife Service (FWS)*

I. The National Wildlife Refuge System (NWRS) is collaborating with EPA, states, local agencies, and USGS to identify watershed-level factors contributing to Clean Water Act 303(d)-listed waters within or near refuges.

II. FWS will continue to coordinate with EPA through a National Memorandum of Arrangement to help ensure that water quality criteria and state water quality standards adequately protect threatened and endangered species.

III. The National Fish Habitat Action Plan (NFHAP) is coordinating with states and communities to protect and restore aquatic habitats.

IV. FWS will continue to conserve clean water through its water pollution prevention activities. These include coordinating with industry on the SmartRx Disposal program for pharmaceuticals and with EPA, U.S. Army Corps of Engineers (USACE), and other agencies on the potential for their actions to affect threatened or endangered species inhabiting, or depending on, water.

d. *Bureau of Indian Affairs (BIA)*

I. BIA will continue to assist tribes in protecting and managing their trust water resources with appropriated funds in the BIA's Water Management, Planning and Pre-development Grant Program.

II. BIA will continue to identify opportunities to protect groundwater basins for tribal, municipal and urban needs such as through DOI/BIA Tribal water rights settlements (e.g., Soboba Water Right Settlement establishing a partnership with water districts to recharge groundwater basins in Southern California).

III. BIA will coordinate with tribes and USGS to identify within the WaterSMART Availability and Use Assessment the basins in which tribes have adjudicated or otherwise enforceable (e.g., congressionally approved settlements) water rights.

IV. BIA will coordinate with tribes and Reclamation to identify basins that are suitable for inclusion in the Basin Study Program.

V. BIA will work with tribes to encourage cooperative studies with other Federal, state, and local agencies to identify adequate water supplies for the future.

e. *Bureau of Land Management (BLM)*

I. As part of its landscape approach for managing public lands, the BLM will conduct Rapid Ecoregional Assessments (REAs) to determine, among other key resource values, areas of regionally significant aquatic habitats. BLM will work with partner agencies and stakeholders to develop broad-level management strategies for the public lands in an ecoregion.

f. National Park Service (NPS)

I. NPS is supporting LCCs and CSC activities through positions within those organizations and providing expertise on scenario planning, cultural resource adaptation, coastal adaptation, landscape ecology, renewable energy, and policy.

II. NPS is participating in the implementation of USGS WaterSMART Availability and Use Assessment with an emphasis on the USGS Colorado River Pilot Study and implementation of the environmental needs component of the WaterSMART

III. Availability and Use Assessment in National Park units.

g. Bureau of Reclamation (Reclamation)

I. Reclamation is coordinating its water conservation activities with partners and stakeholders and expanding existing partnerships through programs such as LCCs, CSCs and, the Basin Study Program. LCCs will include a broad set of missions and views from the Federal, state, tribal and NGO sectors, as well as existing partnerships.

II. Reclamation coordinates with non-Federal partners through WaterSMART Water and Energy Efficiency Grants and its Water Reclamation and Reuse Program (Title XVI).

III. Through the Basin Study Program, Reclamation will conduct West-Wide Climate Risk Assessments to provide identification of risks to water supplies and demands and impacts to operations within the eight major river basins identified in the SECURE Water Act.

IV. Reclamation is responsible for meeting and quarterly reporting on the Secretary's Water Conservation Priority Goal. The Goal gauges progress in implementing actions in partnership with others to conserve water. The goal is - Enable capability to increase available water supply for agricultural, municipal, industrial, and environmental uses in the western United States by 350,000 acre-feet by 2012 through the bureau's conservation-related programs, such as Title XVI and WaterSMART Grants, and an additional 140,000 acre-feet through the end of FY 2012 for a total of 490,000 acre-feet of water savings. Reporting and tracking is coordinated through the Office of Planning and Performance Management in Policy Management and Budget.

h. *Office of Surface Mining Reclamation and Enforcement (OSM)*

I. Pursuant to the Surface Mining Control and Reclamation Act (SMCRA), OSM establishes the minimum Federal standards governing the regulation of coal mining to, among other things, minimize the adverse impacts of mining on streams both within and beyond the SMCRA permit area. With input from FWS, EPA, and USACE, OSM is considering refining its coal mining regulations to better protect streams from the adverse impacts of coal mining, some impacts have only recently been identified by the scientific community. The proposed rule may establish more stringent criteria for mining in or near streams, improve baseline data collection and monitoring requirements, and require better restoration of streams to their pre-mining form and function. The proposed rule would define material damage to the hydrologic balance, which could have the effect of providing increased protection for both ground and surface waters.

II. OSM is continuing its collaboration under an MOU with EPA and USACE to improve coordination among the three agencies, and with state and local governments, regarding regulatory and other activities related to surface coal mining operations governed under SMCRA and the Clean Water Act (CWA).

III. Through its OSM/VISTA community partnerships, OSM-sponsored Watershed Teams are collaborating with local communities affected by past coal mining degradation to improve watershed quality and the environment through water monitoring, water quality project development, water quality training, and environmental education and outreach.

i. *U.S. Geological Survey (USGS)*

I. USGS is coordinating with stakeholders and partners through its implementation of the WaterSMART Availability and Use Assessment (National Water Census) and through involvement in implementing CSCs and LCCs.

II. USGS will also continue to coordinate with other federal agencies, including EPA and USDA on efforts to follow up on water use efficiency recommendations in the October 2010 Interagency Climate Change Adaptation Progress Report.

SECTION 3. THE ENERGY/WATER NEXUS: WATER USED IN ENERGY PRODUCTION AND ENERGY USED IN WATER SUPPLY

Water and energy are inextricably linked. An increased emphasis on domestic energy development will pose additional pressure on limited water supplies. Water is utilized in electric power generation, as well as natural resource extraction and refinement. Significant amounts of water may be required for unconventional and renewable energy development, such as the production of feedstock, and for sequestration of the carbon byproducts of power generation. Energy is utilized in water transmission, treatment, and heating for domestic uses, and for the treatment of wastewater. Section 4(a)(1) of Secretarial Order 3297 directs DOI to identify how much water is used for various energy production technologies and to incorporate such information into decision-making on the development of energy and water resources. DOI coordinating across programs and with other Federal agencies, with states, local governments, tribes, and non-governmental organizations will accomplish this through the following activities.

a. *Department of the Interior (DOI)*
   I. General communication of water footprints to occur through the LCCs, CSCs, and the USGS WaterSMART Availability and Use Assessment.
b. *Fish and Wildlife Service (FWS)*
   I. FWS water programs are examining energy development projects for their potential to adversely affect water rights, groundwater levels, and water quality FWS lands.
   II. ii. Energy audits are conducted on FWS facilities on a 4-year cycle and are used to identify potential savings in the facilities energy budget. Water pumping and conveyance are large energy consumers at many FWS facilities. Improvements to pumping infrastructure can have significant positive impacts on energy costs. Also, reductions in water use can have significant energy savings. Energy audits also identify technologies for both efficient operation and green energy generation. Installation of variable speed drives on motors use only as much wattage as needed for the purpose. Wind power, photo voltaic, geothermal, and micro hydro

turbines generated power are used to offset peak demand or assume full loads at times.

c. *Bureau of Indian Affairs (BIA)*
   I. BIA will look for opportunities, in coordination with affected Indian tribes, to improve conjunctive use of water and energy at BIA's Indian irrigation projects to the extent feasible.
   II. BIA will work with tribes to identify opportunities for saving water and energy through research and in coordination with state and local water resource agencies.
   III. BIA will work with tribes to find opportunities to use geothermal energies such as with the Fort Bidwell Indian Community in Northern California.

d. *Bureau of Land Management (BLM)*
   I. BLM will continue to encourage the sustainable development of water resources and work closely with partners to evaluate efficient uses of the resources while reducing impact to water quality. In the case of energy development through the implementation of best management practices, BLM will continue to protect important water sources and local water quality. BLM will also consider reduced water alternatives for solar thermal energy projects on public lands.

e. *National Park Service (NPS)*
   I. NPS is working with BLM's Programmatic EIS for Solar Energy Development with a major focus on water use and sustainability and protection of park water-dependent resources.
   II. NPS coordinates across parks and with other Federal agencies responsible for leasing and permitting of energy projects on public lands through its service-wide energy team. This team focuses on protection of park water-dependent resources, as well as water quality and use and sustainability of energy development technologies.
   III. NPS has staff co-located with BLM's California Renewable Energy Action Team in Sacramento, California.

f. *Bureau of Reclamation (Reclamation)*
   I. Reclamation's implementation of the SECURE Water Act provisions of P.L. 111-11 will include the Power Resources Office activities under Section 9505 and implementation of the Basin Study Program (including Basin Studies, LCCs, and West-Wide Climate Risk Assessments) under Section 9503.

II. Criteria for WaterSMART System Optimization Review Grants prioritize proposals that evaluate potential renewable energy and energy efficiency improvements in the delivery of water, in addition to identifying options to improve water efficiency.

III. Criteria for the WaterSMART Advanced Water Treatment Grants prioritize projects that consider reduced energy requirements and renewable energy components as new sustainable water supplies are brought online.

IV. WaterSMART Water and Energy Efficiency Grants provide funding for projects that conserve water and projects that address the connection between water use and energy use. For example, through 2010 WaterSMART Grants, the Laguna Madre Water District in Port Isabel, Texas, will make improvements to its non-potable water system, including installation of an energy recovery turbine to create an expected 17,520 kilowatt-hours per year of electricity from wastewater flows. The Lower Colorado River Authority in Texas will automate eleven check gate structures, improvements expected to result in approximately 2,560 acre-feet of water savings annually. The Authority estimates that completion of the project will decrease pumping needs and will reduce energy consumption by approximately 132,000 kilowatt-hours per year.

g. *U.S. Geological Survey (USGS)*

I. USGS will lead development of information and science on amount of water used by different types of electrical generation plants and water required for carbon sequestration.

II. USGS will analyze and collect information on water footprints through the National Water Use Information Program.

III. USGS will participate with coordination activities under GAO Report 10-23 with DOE's Energy Information Administration (EIA). This GAO report calls for USGS to expand its efforts to provide information on alternative water sources used for thermoelectric power generation, such as treated effluent and saline groundwater, to reinstating collection and distribution of water consumption data at thermoelectric power plants, and to improve the overall quality of data collected on water use from power plants by establishing a process for regularly coordinating with the EIA, water and electricity industry experts, environmental groups, academics, and other Federal agencies.

IV. USGS will be working with Reclamation, USACE, DOE, and the Tennessee Valley Authority to assess the water use aspects of hydropower facilities throughout the Nation.

V. USGS will be working with the DOE EIA to develop new methods for surveying and reporting water use-energy use information and how it can be interpreted to provide new insights to energy development.

SECTION 4. BEST AVAILABLE SCIENCE

Under Section 4(a)(2) of the Secretarial Order, the Department will work through the Energy and Climate Change Council, CSCs, and LCCs to obtain best available science to understand the impacts of climate change on water supplies, and to ensure integration of sustainable water strategies within the field offices of DOI's bureaus and agencies. To ensure that the Nation is well prepared to adapt to climate change, it is necessary to evaluate current changes in snowpack, timing and quantity of runoff, changes in groundwater recharge and discharge, environmental conditions, and changes in demands for water. The measurement of available water supplies and changes in those supplies over time, both in surface and ground water, will lead to a better understanding of shortages now and in the future, and which will facilitate the resolution of conflicts over water. Downscaled climate models are necessary, as well as thorough scientific analysis of the historical data. This information must be presented in a format that is useful to the water resources community, to aid in the development of long-term plans flexible enough to adapt to climate change. This section of Secretarial Order 3297 will be implemented through the following activities.

a. *Department of the Interior (DOI)*

I. Bureaus will communicate with CSCs and other programs that are in support of Secretarial priorities as specified with developing the CSCs.

II. The LCCs will develop climate change science focused on landscape conservation decisions and provide a forum and mechanism to facilitate the communication and development of best available science for resource conservation to be shared with partners for bureau-specific and stakeholder-specific activities.

b. *Fish and Wildlife Service (FWS)*

I. The NWRS is implementing a system-wide inventory and assessment of water resources (quantity and quality) in conjunction with NWRS Inventory and Monitoring program. The inventories will establish baselines for NWRS water quantity and quality, and the assessments will identify needs and threats to Refuge System water resources, including threats from climate change. In addition, the assessments will provide information on the efficiency of water use on NWRs. This information will also be important to LCCs for identifying vulnerabilities and assessing future impacts of climate change.
II. The NWRS is conducting hydro-geomorphic (HGM) analyses at multiple National Wildlife Refuges across the country. HGM analyses will identify ecological functions and help the NWRS in the restoration and management of refuge wetlands.
III. NWRS is hiring a full-time water resources coordinator to address national and multi-regional water resource issues in the Refuge System, and to work collaboratively with other bureaus/agencies on projects of mutual concern.
IV. FWS funding will help provide science needed to support LCCs in assessing impacts of climate change on aquatic habitats and DOI trust resources.
V. FWS, through its Fisheries Program and in partnership with EPA, will be using GIS-based overlays to identify water quality issues that may impact National Fish Hatcheries and the imperiled aquatic species propagated and held in refugia at NFHs.
VI. Private landowners own and manage 73 percent of our Nation's land. FWS Partners for Fish and Wildlife Program and the Coastal Program will continue to strengthen partnership efforts at local levels, across the country, to restore and enhance wildlife and coastal habitats. A key component in many of these efforts is restoring and enhancing wetlands, riparian areas and streams to improve water quality and quantity. These Programs also will continue to work with the USDA Farm Bill Programs to provide technical assistance to the design and delivery of national conservation programs administered through the National Resources Conservation Service.

c. *Bureau of Indian Affairs (BIA)*

I. BIA will continue to use the best available science in support of the development and administration of Indian water rights.

II. BIA will seek CSC and regional LCC recommendations, along with utilizing USGS and Reclamation water resource assessment manuals and guidelines, as appropriate, for improving water resource program activities and projects.

III. BIA will explore implementing a climate and water resources monitoring program to provide baseline data for climate change impacts and water resource assessments on Indian trust lands.

d. *Bureau of Land Management (BLM)*

I. A key component of BLM's landscape approach for managing public lands is the completion of the REAs. REAs will establish baseline ecological data to gauge the effect and effectiveness of future management actions on areas of regionally significant aquatic habitats and other areas. In this way, REAs provide a foundation for an adaptive management approach that enables implementation strategies to adjust to new information and changing conditions.

II. As part of its role in the stewardship of the National Hydrographic Dataset, BLM will continue to develop GIS-based tools capable of analyzing a variety of hydrologic characteristics, including locations of impaired streams and estimation of flow from contributing sources.

III. BLM's Assessment, Inventory, and Monitoring (AIM) Strategy will standardize data collection and retrieval so information is comparable over time, and can be readily accessed and shared. The goal of the AIM Strategy is to provide the information needed to understand trends in resource conditions from climate change, including water quantity and quality, and to evaluate and refine implementation actions.

IV. BLM is implementing its Geospatial Services Strategic Plan.

e. *National Park Service (NPS)*

I. NPS will engage in short- and long-term water resources monitoring and science in support of the protection of park water resources. For example, NPS provides support to and participates in interagency basin-wide assessments of aquifer sustainable yields in water scarce areas such as the Great Basin and southern Nevada, and conducts site-specific aquifer drawdown assessments of proposed water developments to determine the potential for impacts to park water resources. NPS also conducts and supports studies to quantify instream flow needs for flow-dependent park resources potentially influenced by water developments or the operation of water storage projects.

II. The NPS Inventory and Monitoring program engages in the acquisition of scientifically sound information obtained through systems-based inventory and monitoring (I&M). NPS inventories park natural resources to determine their nature and status and conducts long-term ecological monitoring, known as Vital Signs Monitoring, to better understand the dynamic nature and condition of park resources and to provide reference points for comparisons with other altered environments. NPS conducts park Vital Signs Monitoring through 32 eco-regional networks covering 270 parks with significant natural resources. The information obtained through the I&M program has multiple applications for management decision-making, park planning, research, education, stakeholder involvement, and promoting public understanding of park resources.

III. NPS will pursue enhanced or expanded monitoring of climate-sensitive indicators by the NPS Vital Signs Monitoring Networks to better understand the extent, rate, and magnitude of climate effects on park resources, including park water resources.

f. *Bureau of Reclamation (Reclamation)*

I. Reclamation's WaterSMART Grants to develop Climate Analysis Tools will provide funding to improve the ability to evaluate the impacts of climate change on water resources.

II. Reclamation's Research and Development Science and Technology Program activities will be shared and coordinated to complement activities by the CSCs and USGS to support Secretarial priorities.

III. In implementing the Basin Study Program (including Basin Studies, LCCs, and West-Wide Risk Climate Assessments), Reclamation collaborates and shares information with LCC partners, other Federal agencies, including USGS, NOAA, NOAARegional Integrated Sciences & Assessments (RISAs), and USACE, as well as non-federal stakeholders, to evaluate the risks and impacts of climate change to water supply, demand and energy production. This is consistent with Section 9503 of the SECURE Water Act, which also mandates that Reclamation report to Congress every five years on these factors and the adaptation options identified.

IV. Reclamation will continue to utilize the best available science to maintain infrastructure for a sustainable water management system.

g. *U.S. Geological Survey (USGS)*

I. USGS will support CSCs. CSCs will provide science support to LCCs to meet identified needs related to the assessment of climate change impacts to natural resources.

II. The National Water Use Information Program, which will be an integral part of the Water Availability and Use Assessment, will provide more accurate assessments of water resources in terms of past and current supplies and uses and how these factors result in conflicts and shortages. This is consistent with Section 9508 (d)(5) of the SECURE Water Act, which also mandates that USGS report to Congress every five years on the factors responsible for these conflicts and shortages. This information will assist in shaping public policy to reduce conflicts in water management. Water Data Enhancement through the National Streamflow Information Program, as reviewed by the National Research Council in 2004, will ensure that the streamflow and related environmental variables throughout the Nation are collecting the data necessary to facility water resource management and better understanding of hydrologic extremes. This is consistent with Section 9507(a) of the SECURE Water Act.

III. USGS will gather and dissemination data from Landsat thermal infrared sensor to assist in understanding national moisture and drought conditions.

IV. USGS will work with USACE through the ACWI Subcommittee on Sedimentation to catalog rates of sedimentation within selected reservoirs of the Nation.

SECTION 5. WATER FOOTPRINT REDUCTION PROGRAM

Section 4(a)(3) of the Secretarial Order calls for a water footprint reduction program for DOI facilities and water-consuming operations. Accordingly, DOI will provide input and information on water conservation methods and technology to bureaus and offices and will work with the Senior Sustainability Officer to achieve and exceed goals set forth in Executive Order 13514. DOI will implement this section through the following activities.

a. *Department of the Interior (DOI)*

   I. Assistant Secretary Policy, Management, and Budget is leading the implementation of DOI's Strategic Sustainability Performance Plan (SSPP), as authorized by Executive Order 13514 – Federal Leadership in Environmental, Energy, and Economic Performance. The SSPP provides a roadmap for managing and reducing the Department's environmental impacts while meeting our mission responsibilities.

   II. DOI has implemented a Department-level environmental management system (EMS) to manage and track progress on achieving the goals of the SSPP. The foundation of the EMS is the Department's Sustainability Council. The Council is the implementing and oversight body for the EMS and SSPP and is responsible for coordinating and integrating departmental programs and staff to achieve the goals of the SSPP.

b. *Fish and Wildlife Service (FWS)*

   I. FWS is tracking, reporting, and reducing potable water use per the requirements of Executive Order 13423, including potable water used for landscape irrigation. For the second consecutive year, the Director has issued a memorandum to approximately 125 field stations reminding them to achieve water use reductions and to report potable (domestic) water consumption (as opposed to production water use). This memorandum emphasizes that full compliance with Executive Order 13423 will only be achieved by making radical changes in our operations and maintenance practices, and includes technical guidance on how to conserve water in buildings and landscaping.

   II. FWS will continue to pursue water conservation activities including through emphasizing water reuse at National Fish Hatcheries, identifying infrastructure investments that will increase efficiency of refuge water use.

c. *Bureau of Indian Affairs (BIA)*

   I. BIA will work with tribes and other Federal, state, and local agencies to identify opportunities to reduce water use in activities and programs.

d. *Bureau of Land Management (BLM)*

   I. BLM's draft Strategic Plan for Integrating Greening Activities includes the Energy and Water Conservation goal to reduce water consumption intensity 16 percent by the end of FY 2015 from baseline of FY 2007.

e. *National Park Service (NPS)*
   I. Green Parks Plan: To contribute to the responsible use of fresh water supplies and preserve fresh water in-flows in rivers and streams, as well as groundwater supplies, the NPS will demonstrate leadership in water conservation, gray water reuse, rainwater capture, and the reduction of wastewater conveyance. NPS intends to set a target of a 3% annual reduction in Service-wide water consumption and a 30% reduction in wastewater generation by 2020. NPS will also set a target to reduce stormwater runoff from existing buildings and new construction/major renovation sites. NPS expects to finalize the Green Parks Plan in 2011.

f. *Bureau of Reclamation (Reclamation)*
   I. Reclamation's development of the Strategic Sustainability Performance Plan and implementation through EMS activities and Energy scorecard.

g. *U.S. Geological Survey (USGS)*
   I. USGS, through the National Water Use Information Program, will provide information on water use and trends in water use by the various sectors it tracks and reports.
   II. USGS will work with the Federal Energy Management Program's Interagency Water Working Group to track progress on achieving the water footprint reduction goals laid out in Executive Order 13514.

SECTION 6. WATERSMART CLEARINGHOUSE

Section 4(b) of the Secretarial Order directs the establishment of a WaterSMART Clearinghouse to provide information to the public through a website which will include information on best practices case studies, education and outreach, financial assistance, laws, regulations, policies, research, and water data and information. This information will allow for better coordination at the Federal level but also support state and local governments, tribal nations, and others in coordinating and integrating water conservation and sustainable water strategies. DOI's activities to accomplish this section are as follows.

a. Advisory Committee on Water Information (ACWI). At the annual meeting of ACWI on July 13 – 14, 2010, the Assistant Secretary for Water and Science requested ACWI's participation in reviewing and improving the WaterSMART Clearinghouse Website. ACWI's Sustainable Water Resources Roundtable (SWRR) has assembled a workgroup of members from throughout the ACWI organization to review the Clearinghouse and has provided recommendations to DOI.
b. Fish and Wildlife Service (FWS)
   I. FWS will contribute information for the WaterSMART clearinghouse including case studies, best practices, conservation activities and other information on how the agency is conserving water resources.
c. Bureau of Indian Affairs (BIA)
   I. BIA will regularly review and, as appropriate, release relevant, and adjudicated trust water resource data to the WaterSMART Clearinghouse.
d. Bureau of Land Management (BLM) i. BLM will contribute information for the WaterSMART clearinghouse relevant to sustainability of water resources and its stewardship responsibilities.
e. Bureau of Reclamation
   I. Reclamation will work in the development of the WaterSMART Clearinghouse; provide information on conservation programs and activities, case studies, and best practices.
f. U.S. Geological Survey (USGS)
   I. USGS will work in the development of the WaterSMART Clearinghouse; Clearinghouse to represent information on water conservation information, science, and tools from bureaus.

SECTION 7. PROMOTING SUSTAINABLE WATER STRATEGIES

Section 4(c) of the Secretarial Order calls for DOI to apply criteria that promote sustainable water strategies and development of incremental hydroelectric power generation.

This section of Secretarial Order 3297 will be implemented through the following activities.

a. *Fish and Wildlife Service (FWS)*
   I. FWS will coordinate with Reclamation and USGS to identify sustainable strategies including healthy aquatic habitats and the DOI trust resources that depend on them.
   II. The NWRS will continue to use instream flow protections to maintain sustainable water supplies for fish, wildlife, and habitats.
   III. The NWRS water resources inventories and assessments will help lead to sustainable water strategies by identifying and promoting efficient water use.
   IV. FWS will protect water rights. Water rights (especially in the West) must be protected and maintained in order to sustain the habitat and species that rely on an adequate water supply for our National Wildlife Refuges and Fish Hatcheries.
   V. FWS will maintain National Pollution Discharge Elimination System (NPDES) permits at National Fish Hatcheries to demonstrate FWS's commitment to protecting water for fish and wildlife as well as for downstream users.

b. *Bureau of Indian Affairs (BIA)*
   I. BIA will coordinate with tribes and USGS to identify within the WaterSMART Availability and Use Assessment the basins in which tribes have adjudicated or otherwise enforceable (e.g., congressionally approved settlements) water rights, and what those water rights are.
   II. BIA will coordinate with tribes and USGS to identify within the WaterSMART Availability and Use Assessment basins in which considerable conflict exists between existing water uses and tribal trust or treaty resources, and the tribal water rights have not yet been adjudicated.

c. *Bureau of Land Management (BLM)*
   I. BLM will continue to participate in, and provide funding for, ACWI's SWRR in support of its mission to promote sustainability of the nation's water resources through the evaluation of information, development and use of indicators, targeting of research, and the engagement of people and partners to improve the management, conservation, and use of water and related resources.

d. *Bureau of Reclamation (Reclamation)*
   I. Reclamation will provide WaterSMART Grants (Water and Energy Efficiency, System Optimization Review, Advanced Water

Treatment Pilot and Demonstration, and Research) to support sustainable water resources in the Western United States.

II. Reclamation will continue to implement the Water Reclamation and Reuse Program (Title XVI) program. Title XVI projects require coordination and compliance with all applicable Federal, state and local laws.

III. Reclamation will continue to implement the Water Conservation Field Services Program (WCFSP) to encourage water conservation and efficient use of water supplies on Reclamation's projects and to foster improved water management on a watershed basis throughout the western states.

IV. Reclamation is working with USACE on the Hydropower Modernization Initiative to assess the potential for increased capacity and efficiency gains at its existing hydroelectric plants.

V. Reclamation is conducting an operations optimization study that will implement hydropower dispatch optimization programs at Reclamation's hydropower plants.

VI. Reclamation is participating in a Memorandum of Understanding for Hydropower with DOE and DOA (through USACE). The MOU parties will focus both on increasing renewable energy generation from Federal hydropower facilities and reducing the environmental impact sometimes associated with historical hydropower development in the U.S. by focusing on sustainable, low impact and small hydropower projects and collaborate on efforts to implement projects at these locations, including joint studies, demonstration projects and other mutually supported partnership arrangements with private entities, Indian tribes, and state and Federal agencies.

VII. vii. Reclamation is performing an updated 1834 resource assessment a draft of which was completed in October 2010.

e. *U.S. Geological Survey (USGS)*

I. USGS to identify programs and activities related to promoting sustainable water strategies. One way to promote sustainable water strategies is through thorough characterization of the water resources and the stresses influence that resource. USGS can assist by careful long-term monitoring of the major components of the hydrologic cycle and by quantifying the degree of uncertainty associated with those measurements. USGS will strive to do this through its monitoring programs.

SECTION 8. EVALUATION OF NEEDED INFORMATION

Under Section 4(d) of the Secretarial Order, the WaterSMART Program will include evaluation of information needed for sound decision-making on sustainable water, along with a determination as to whether existing information about the availability, quality, and use of water is adequate or requires enhancement.

The evaluation of needed information will be done through the coordination with other Federal agencies, state, tribal, and local governments as well as non-governmental organizations. Specific activities to meet this portion of Secretarial Order 3297 are identified as follows.

a. *Department of the Interior (DOI)*
   I. All Bureaus participating in providing to and using information from the USGS WaterSMART Availability and Use Assessment.
b. *Fish and Wildlife Service (FWS)*
   I. FWS will coordinate with LCCs and specifically with USGS to identify science and data needs for sustaining healthy aquatic habitats and DOI trust resources that depend on them.
   II. The NWRS System-wide inventory and assessment of water resources (quantity and quality) will provide information on the efficiency of water use on NWRS and provide the Refuge System with needed information to prioritize water resource needs.
c. *Bureau of Indian Affairs (BIA)*
   I. BIA will evaluate on an annual basis, in conjunction with the Indian water rights negotiation and litigation teams, technical information needed to support sound decision-making for the protection and development of Indian water rights and resources.
   II. BIA will facilitate outreach and education to local communities to address tribal water resources concerns and shared water resources needs within the watershed regions.
d. *Bureau of Reclamation (Reclamation)*
   I. Reclamation can use information provided through the USGS WaterSMART Availability and Use Assessment to inform water demand projections for project specific work, Basin Studies, and West-Wide Climate Risk Assessments.
   II. Reclamation and its LCC partners will assess what information gaps exist with the LCCs and will attempt to address those gaps by

working with LCC partners, including USGS and CSCs, and universities and research institutes, among others.

III. Reclamation, with USGS, NOAA, EPA, and USACE continues to develop an ongoing needs assessment, updated in 2010, to identify the current and future science related to climate change information to inform climate change planning decisions. This activity builds upon USGS Circular 1331 published in 2009.

e. *National Park Service*

I. NPS will work with others to enhance the application of hydrologic modeling to help bracket potential alternative water availability futures for parks. NPS will apply climate, surface and ground water models necessary to assess influences of climate change on precipitation, runoff, and aquifer recharge, and to help evaluate alternative water development proposals, reservoir operations, and allocation scenarios.

II. NPS will continue to work with other agencies on actions that may affect water available to park resources including surface water withdrawals, reservoir operations plans, ground water development, Clean Water Act sections 401 and 404 permits, and endangered species considerations.

III. NPS will strengthen its understanding of the relationships between water availability and environmental needs. NPS will identify water dependent resources in parks that are vulnerable to the effects of increased water demand, or climate changes and initiate studies to determine their instream flow and/or ground water needs.

f. *U.S. Geological Survey (USGS)*

I. USGS will implement the WaterSMART Availability and Use Assessment

II. Complementary to the above assessment, the USGS with Reclamation, NOAA, EPA, and USACE will identify the best ways to develop the science and data necessary to meet the water management needs to support sound decision-making on water conservation and sustainability.

This analysis will be done in concert with the activities under Section 9506 of the SECURE Water Act and will be available for review by April 2011.

SECTION 9. EDUCATION AND AWARENESS

Section 5(b) of Secretarial Order 3297 directs that appropriate education, awareness, and outreach actions be taken. These actions will be coordinated with other Federal, state, tribal, and local governments as well as non-governmental organizations. The following activities will be undertaken or continued to meet this section of the Secretarial Order.

a. *Department of the Interior (DOI)*
   I. An outreach plan will be developed that will include talking points, presentation materials, fact sheets, and other necessary and commonly used briefing material for internal and external communications.
   II. The WaterSMART Clearinghouse as identified in Section 6 is also a component of Education and Awareness.
   III. DOI's Sustainability Council is responsible for promoting and achieving departmental goals for sustainability and compliance. This includes the internal dissemination of information regarding water use efficiency and management of their bureau program and field offices.

b. *Fish and Wildlife Service (FWS)*
   I. FWS will feature WaterSMART information, case studies, and best practices using existing communications tools such as the website, news releases, speeches, and internal communications tools.

c. *Bureau of Indian Affairs (BIA)*
   I. BIA will promote education and training for tribal members sponsored by their tribe to participate in the Bureau's Water Resources Technician Training Program.

d. *Bureau of Land Management (BLM)*
   I. BLM will increase its outreach efforts through the development of "At-A-Glance" factsheets for each of its water-emphasis programs: Surface Water, Water Rights, Water Quality, and Groundwater. In part, these factsheets will relay BLM success stories in areas of water sustainability

e. *Bureau of Reclamation (Reclamation)*
   I. Reclamation conducts education and awareness activities through the WCFSP in some regions.

f. *National Park Service (NPS)*

I. NPS will advocate for water education by proposing a partnership project with Project WET called "Discover the Waters of Our National Parks."

g. *Office of Surface Mining Reclamation and Enforcement (OSM)*

I. OSM will continue outreach and education through its OSM/VISTA partnerships with states and local communities to promote water quality improvements to address past coal mining degradation.

h. *U.S. Geological Survey*

I. The USGS will continue to promote water education and awareness through its on-going programs in "Water Basics" (http://water.usgs.gov/education.html), its distribution of selected products of educational interest, its support of multimedia resources for water education, and its support of internships and research grants. USGS maintains online science education resource materials at: http://education.usgs.gov/

SECTION 10. COLLABORATION WITH STATES AND TRIBES

Section 5(c) of Secretarial Order 3297 provides that DOI will work within existing relationships, and develop new partnerships, with states and tribes to collaborate on implementation of the WaterSMART strategy.

a. *Department of the Interior (DOI)*

I. LCCs can facilitate the development of new partnerships with states and tribes to collaborate on WaterSMART implementation and strategy.

b. *Fish and Wildlife Service (FWS)*

I. FWS will collaborate with states and tribes within the LCC network.

II. FWS Tribal grants will consider climate change effects on aquatic resources.

III. The National Fish Habitat Action Plan (NFHAP) is a collaborative effort among many organizations, including FWS, states and tribes, to protect and restore aquatic habitats.

c. *Bureau of Indian Affairs (BIA)*

I. BIA will continue coordination with tribes as well as state and local governments in the development of water resources needs and plans through the use of cooperative agreements, Integrated Regional Water Management programs, and promotion of Indian water right requirements and perspectives.

II. BIA will identify alternative funding sources with Federal, state and local agencies to study shared watersheds (e.g., Cooperative Watershed Management Program authorized by P.L. 111-11) and assist tribes in developing cooperative agreements for shared management of local resources that include both state and Federal concerns.

III. BIA will look for opportunities in the context of Indian water rights negotiations where tribes and the non-Indian parties can consider water supply projects which promote sustainability and conservation of water sources as part of Indian water settlements.

IV. BIA will encourage follow up actions on the results of a Tribal Water Summit held in November 2009 with the California Department of Water Resources, tribal governments and federal agencies to obtain tribal comments on California's Water Plan 2009. One of the outcomes of the meeting was a list of recommended actions to address tribal water issues within California.

d. *Bureau of Land Management (BLM)*

I. BLM will continue to participate in the WestFAST consortium and the Western States Water Council on sustainable water use and development issues.

e. *Bureau of Reclamation (Reclamation)*

I. Desert and Southern Rockies LCCs can serve to develop new partnerships with states and tribes and collaborate on WaterSMART implementation and strategy.

II. Reclamation also collaborates with partners and creates new partnerships through WaterSMART Grants, Basin Studies, Title XVI and WCFSP.

f. *National Park Service (NPS)*

I. NPS participates in state water court and administrative processes to protect park water-dependent resources and values. NPS played a key role in the development of the precedent-setting Water Rights Compact between the state of Montana and the United States. The compact resolved NPS Federal-reserved water rights claims in six national parks in Montana and protected the hydrothermal features of Yellowstone National Park. NPS works through state administrative processes to protect park water-dependent resources from water development proposals such as the large-scale groundwater development in southern and east-central Nevada, and conventional and renewable energy projects. NPS focuses on the acquisition and application of high-quality science, including development of regional-scale groundwater models, and effective communication with stakeholders in addressing water development proposals that could impact park resources.

II. NPS will coordinate with tribes to address water allocation on a basin-wide scale.

III. NPS works with states to protect and preserve high quality waters in units of the National Park System through Clean Water Act designations such as Outstanding National Resource Waters.

g. *Office of Surface Mining Reclamation and Enforcement (OSM)*

I. Once its Stream Protection Rule is finalized, OSM will work with state and tribal regulatory authorities to educate them on the changes they must make to their programs. OSM will also provide training to help local regulators understand and implement the new measures to provide additional protection for groundwater and surface water.

h. *U.S. Geological Survey (USGS)*

I. USGS will work through CSCs to develop new partnerships with state governments, universities, tribes, NGOs, etc.

II. USGS will work through its Cooperative Water Program partners, which include both states and tribes, to explore opportunities to further the requirements of WaterSMART.

III. USGS will develop new strategies and science consistent with WaterSMART goals.

SECTION 11. PLANNING EFFORTS

Section 5(d) of Secretarial Order 3297 calls for recommendations to incorporate water conservation criteria and the water/energy nexus into DOI's planning efforts, including recommendations to reduce conflict in water management. Planning efforts will be coordinated with other Federal, state, tribal, and local governments as well as nongovernmental organizations. The activities that will accomplish this are listed below.

a. *Department of the Interior (DOI)*

   I. DOI's's Asset Management Program requires that all building projects, regardless of type, must follow all applicable mandatory energy, water, and sustainable buildings requirements within the scope of the project. Specific requirements for new and existing buildings are provided in the DOI Sustainable Buildings Assessment and Compliance Tool.

b. *Fish and Wildlife Service (FWS)*

   I. FWS through the NWRS is implementing a System-wide inventory and assessment of water resources (quantity and quality) in conjunction with NWRS Inventory and Monitoring program that will help prioritize needs to obtain, protect, and manage water resource quantity and quality in the NWRS.

   II. FWS has completed a preliminary analysis of climate change water vulnerabilities in FY09, and identified vulnerable Refuges and Fish Hatcheries in each Region. Eastern and western water workgroups put forward two Refuges and a Hatchery from each Team as potential projects.

c. *Bureau of Reclamation (Reclamation)*

   I. Reclamation will explore where planning activities could include water/energy nexus information. For example, water/energy nexus criteria have been included in WaterSMART Grant criteria and Title XVI criteria. Reclamation has also incorporated an emphasis on the water/energy nexus into its Rural Water Program, which provides financial assistance to plan rural water projects.

   II. Reclamation through its Basin Studies will evaluate the impacts of climate change to water and power facilities, including environmental response and natural hazards and will identify mitigation and adaptation strategies that include water conservation, and will look to the development of feasibility studies that incorporate these options.

d. *U.S. Geological Survey (USGS)*

   I. Through the WaterSMART Availability and Use Assessment Geographic Focus Studies, USGS will be looking to identify areas of the Nation where there is significant conflict over water resources or water shortage issues. This is consistent with Section 9508 (d) (5) of the SECURE Water Act, which also mandates that USGS report to Congress every five years on the factors responsible for these conflicts or shortages. This information will assist in shaping public policy to hopefully reduce conflicts in water management.

SECTION 12. COLORADO RIVER BASIN PILOT

Perhaps nowhere is a single water resource more important to a varied user community, (including many states, tribes, fish and wildlife) than the Colorado River. As an area to focus WaterSMART efforts to sustain and stretch this water resource for people, the economy, and the environment, DOI is piloting many efforts within this shared resource. The Department of the Interior recognizes the diversity and significance of all Federal, state, tribal, local, and non-governmental activities along the Colorado River Basin and will coordinate with all of these activities. Within the Colorado River Basin, DOI will have the following specific activities:

Program Coordination

I. Reclamation is co-lead with FWS in the effort to establish two LCCs, the Desert and Southern Rockies and will develop shared science needs focused on the Colorado River Basin. This activity will be coordinated with the ongoing Basin Study in the Colorado River Basin and current and potential future Title XVI activities.

II. The USGS will coordinate across WaterSMART activities within the Colorado River Basin, including a focus area study within the WaterSMART Availability and Use Assessment.

III. NPS is participating on the Colorado River Basin Study with Reclamation and within the USGS Water Availability and Use Assessment.

IV. To improve internal and external communication and coordination, NPS formed a Colorado River Steering Committee comprised of superintendents of Colorado River parks and pertinent regional directors and Washington staff. NPS has become fully engaged in communicating and coordinating its interests in Colorado River management across DOI bureaus through a variety of forums including regular meetings of senior leadership at the regional and Washington levels, participation in development of the Colorado

River Annual Operating Plan, and participation on the Glen Canyon Dam Adaptive Management Workgroup.

V. BLM is coordinating efforts on their Colorado Plateau Ecoregional Assessment with Reclamation's Colorado River Basin Study. These efforts including policy, objective, and methodology discussions and to integrate data sharing.

VI. BLM's Salinity Control Program is participating in a remote-sensing, interagency study with the USGS Center for EROS, Pacific Northwest National Laboratory, and others that will provide vegetational health information with climate-changing conditions. Water availability for vegetation determines phonological-stage temporal and spatial progressions. The study is creating a model to "now-cast" rangeland health, identify production anomalies, quantify effects of different management treatments and determine growth curves of certain critical plant species.

Water Used in Energy Production and the Energy Used in Water Delivery

I. The USGS WaterSMART Availability and Use Assessment will have a regional focus area study focused on the Colorado River Basin. The assessment will include uses of water within the basin for energy production and energy used for water delivery.

II. Reclamation's WaterSMART Water and Energy Efficiency Grants will provide funding for projects that address the connection between water use and energy use. As an example the Eastern Municipal Water District in Southern California will construct a facility to treat water that is currently discharged to the sewer so that additional water can be provided to the Perris Water Filtration Plant for ultimate use as potable water. Once completed, the project is expected to save approximately 950 acre-feet of water each year, resulting in a decrease in imported raw water consumption by the District. The recovery facility is also projected to result in savings of 4,412 kilowatt-hours of electricity for every 1 million gallons of water diverted from the sewer.

III. BLM will review NEPA analysis for energy projects on public lands,

IV. including renewable energy projects, to ensure best management practices and alternatives are considered to minimize impacts to water resources and water quality.

Best Available Science

I. The Desert and Southern Rockies LCCs will convene water resources specific subgroup focused on the science needs of the Colorado River Basin. The first scoping meeting of this subgroup was held in September 2010 in Las Vegas, Nevada
II. The Desert and Southern Rockies LCC will work with the Southwest CSC to identify regional scale science information needed to support the water resources management decisions within the Colorado River Basin.
III. The USGS WaterSMART Availability and Use Assessment will have a regional focus area study on the Colorado River Basin to develop the necessary science to inform current and future water resources decisions.
IV. NPS uses best available science to support all natural and cultural resource management actions in the Colorado River Basin, including actions
V. supporting species recovery, riparian restoration, exotic species control, and the management of recreational use.
VI. BLM's AIM effort to develop a framework structure for scientific resource (water, air, soil, vegetation) information will house historical and ongoing science-project data from the Upper Colorado River Basin Salinity Control Program.
VII. USGS National Streamflow Information Program will provide necessary information within the Colorado River Basin.
VIII. USGS working with the ACWI Subcommittee on Sedimentation are evaluating effects of sedimentation in reservoirs within the Colorado River Basin.

Promoting Sustainable Water Strategies

I. Reclamation through the WaterSMART Water and Energy Efficiency Grants will provide funding for projects focused on sustainable water strategies within the Colorado River Basin. Additionally adaptation options will be explored through the Colorado River Basin through Basin Study and pilot assessment by Reclamation and the USGS, respectively.

II. Through its National Fish Hatcheries, FWS will continue to employ reuse and recirculation strategies to rear native and mitigation sport fish. Additionally net pens will continue deployment in the Colorado River to further reduces electrical demands and pumping. Other extremely cost-effective measures also will be employed, including covering hatchery raceways to reduce evaporation rates in rearing units, which allows more spring water to leave hatchery property and flow to the Colorado River.

Evaluation of Needed Information

I. The Desert and Southern Rockies LCCs will convene water resources specific subgroup focused on the science needs of the Colorado River Basin. The first scoping meeting of this subgroup was held in September 2010 in Las Vegas, Nevada. One focus of the meeting was to identify needed information that can be addressed with science organizations including the Southwest CSC and the NOAA-RISAs and USGS.

Education and Awareness

I. The Desert and Southern Rockies LCCs, through the water management subgroup, will plan to hold science dissemination and collaboration workshops that are focused on establishing the science priorities within the Colorado River Basin and sharing the science that has already been developed.

Coordination with States and Tribes

I. Reclamation's Basin Study and the USGS assessment within the Colorado River Basin will actively encourage and outreach to states and tribes.

II. NPS has a full time Colorado River parks coordinator to interact and coordinate with other Federal agencies (e.g., Reclamation, FWS), various organizations (e.g., Glen Canyon Dam Adaptive Management Workgroup), and tribes.

III. FWS will continue to explore and utilize novel strategies in close consultation and coordination with its Tribal partners, including using hatchery effluent to fill Colorado River Indian Tribes fishing ponds. Effluent is also used on acreages to provide upland and water fowl hunting habitats, which reduces both water demands on the Colorado River and associated pumping costs.

Planning Efforts

I. Reclamation's Basin Study is a cost shared study of climate change water supply and demand risks and impacts study on the Colorado River Basin. The study includes major stakeholders and will explore adaptation options under climate change scenarios.

II. FWS will be implementing risk analyses on its National Fish Hatcheries to determine potential water issues and other factors that could impact work on imperiled aquatic species within the Colorado River Basin.

LIST OF ACRONYMS

ACWI	Advisory Committee on Water Information
AIM	Assessment, Inventory, and Monitoring
BIA	Bureau of Indian Affairs
BLM	Bureau of Land Management
CSC	Climate Science Centers
CWA	Clean Water Act
DOA	Department of the Army
DOE	Department of Energy
DOI	Department of the Interior
EMS	Environmental Management System
EPA	Environmental Protection Agency
EROS	Earth Resources Observations and Science
FWS	Fish and Wildlife Service

GIS	Geographic Information System
HGM	Hydro-Geomorphic
LCC	Landscape Conservation Cooperatives
MOU	Memorandum of Understanding\
OSM	Office of Surface Mining Reclamation and Enforcement
NFHAP	The National Fish Habitat Action Plan
NGO	Non-Governmental Organization
NOAA	National Oceanic and Atmospheric Administration
NPDES	National Pollution Discharge Elimination System
NPS	National Park Service
NWRS	National Wildlife Refuge System
REA	Rapid Ecoregional Assessment
RISA	Regional Integrated Sciences & Assessments
SECURE	Science and Engineering to Comprehensively Understand & Responsibly Enhance
SMCRA	Surface Mining Control and Reclamation Act
SSPP	Strategic Sustainability Performance Plan
SWRR	Sustainable Water Resources Roundtable
Title XVI	Reclamation Wastewater and Groundwater Study and Facilities Act (Title XVI of P.L. 102-575), as amended – Reclamation Water Reclamation and Reuse Program
USACE	U.S. Army Corps of Engineers
USDA	U.S. Department of Agriculture
USGS	U.S. Geological Survey
WCFSP	Water Conservation Field Services Program
WestFAST	Western States Federal Agency Support Team

In: National Water Sustainability
Editor: Willliam Owens
ISBN: 978-1-62618-103-8

Chapter 2

WATERSMART: A THREE-YEAR PROGRESS REPORT*

U.S. Department of the Interior

OVERVIEW

The health, security, and economic well-being of the American people depend on adequate supplies of clean water. We use water to drink and stay healthy, produce energy, grow food, advance the economy, and engage in recreation. On a more basic level, water is vital to maintaining the healthy ecosystems that support our lives. But the amount of water available to support people, the economy, and the environment is finite. As our need for water continues to grow, available supplies will be increasingly strained, risking the depletion of our shared natural resources and raising the potential for conflict.

Three years ago, Congress passed the SECURE Water Act of 2009, directing the United States Department of the Interior to develop a sustainable water management policy. In 2010, Secretary of the Interior Ken Salazar established WaterSMART, combining existing programs with new initiatives to create a broad framework for wisely managing the Nation's water supplies. Through these programs, Interior is actively working with Tribal, State,

* This is an edited, reformatted and augmented version of the U.S. Department of the Interior, dated October 2012.

* This is an edited, reformatted and augmented version of the U.S. Department of the Interior, dated October 2012.

regional, and local water managers to address a range of urgent issues associated with water scarcity.

Sustainable Water Management

Sustainable water management means striving for balance – balance between the water we have, the water we need, and the water we use; between water for the present and water for the future; and between the water we take from rivers, lakes, and aquifers, and the water we leave to preserve the healthy ecosystems on which we depend.

Several factors contribute to the urgent need for balancing water supplies, including population growth and movement, changes in the economy, advances in technology, the movement toward energy independence, and variation in climatic conditions with associated impacts on water. Finite resources coupled with growing demands results in shortage. Continued increases in the demand for all uses of water will place even greater stress on our water supplies, especially the more limited water supplies of the Nation's arid regions.

The evidence of water shortage includes depleted groundwater sources, lower reservoir levels, and reduced stream flows. In April 2005, for example, the water stored behind Glen Canyon Dam in the Lake Powell reservoir was one-third of the full capacity, the lowest level in decades, due to prolonged drought in the Colorado River basin. As a result, not only is the water supply strained, but hydropower generation drops as well. Projected changes in the climate threaten to magnify the imbalance between water supply and water demand. Solving the problems caused by water shortage requires coordinated actions among all levels of government and among public, private, and non-governmental organizations.

SECURE: Science and Engineering to Comprehensively Understand and Responsibly Enhance

WaterSMART: Sustain and Manage America's Resources for Tomorrow

Through the WaterSMART Program, Interior is providing leadership on the path to a secure and sustainable water future because without action, our demand for more water will quickly outstrip the amount available to us. The WaterSMART Program provides the scientific and financial tools and the collaborative environment needed to help balance water supply and demand through the efficient use of current supplies and the development of new supplies. By working together, we can sustain our shared water resources for future generations.

Water stored behind Glen Canyon Dam in the Lake Powell reservoir is at its lowest level in decades.

Improving Water Management through Science

Good resource management relies on accurate and complete information. We cannot successfully manage our water resources without understanding water supply, its many sources and timing, current demands for all uses, and future trends, risks, and uncertainties. Through WaterSMART, Interior is making use of the best available science in the assessments it conducts and the policies it employs. WaterSMART science has and will continue to inform the real-time decisions of water managers who need reliable estimates of current conditions in the hydrologic cycle and projections of supply and demand in watersheds throughout the Nation. Many examples of best available science are being developed through the WaterSMART Program. Much of that science can be accessed through the WaterSMART Clearinghouse, an online collaborative site where best practices and cost-effective technologies for water conservation and sustainable water strategies are shared with the public. Science also underlies the prioritized allocation of grants to support water conservation and sustainability.

Interior delivers the results of its scientific work through a variety of online tools and publications. As described in the chapter, "National Water Availability and Use Assessment," the U.S. Geological Survey initiated the first National Water Census in over three decades. The Census will provide State and local water managers essential information on hydrology and its role in watershed ecology. The USGS also published techniques for estimating water use and evapotranspiration and for assessing ecological water needs. The Bureau of Reclamation produced an online streamflow projection tool for water managers based on published results of collaboratively developed downscaled climate and hydrology projections.

Interior has also established institutions to enhance collaboration on basic and applied science. The USGS National Climate Change and Wildlife Science Center serves the scientific needs of natural and cultural resource managers as they plan for a changing climate. The center manages Interior's Climate Science Centers, which enable the USGS, water managers, and other basin stakeholders to exchange information on the basic science of climate change impacts. Basic science includes physical and biological research, ecological forecasting, and multi-scale modeling. The basic science produced through the Climate Science Centers benefits the members of Landscape Conservation Cooperatives, another collaborative institution described below and later in this report.

By using the best available science, Interior accomplishes two important water resource assessment and management objectives. First, it provides information that is difficult for individual water managers to obtain, with the highest degree of certainty in what is often an uncertain world. Second, it continues to push the goal higher for better management of precious water resources. Federal, State, and local government managers and water resource associations all recognize that they need more and better water data and improved water science – an important aspect of which is the need for improved analytical approaches that will contribute to new, adaptive strategies for management.

Improving Water Management through Collaboration and Cooperation

Interior is the Nation's largest wholesaler of water, steward of one-fifth of the Nation's land, and administrator of the United States' trust responsibilities to Native Americans. Our understanding of water resource management has expanded from simply supplying water for domestic and irrigation needs to providing water for ecosystem health, municipal and industrial demands, energy and economic development, and recreation. At the same time, multiple levels of government and non-governmental organizations are involved. State laws govern the use, allocations, and management of most water resources and local laws govern many land use decisions. Communities are increasingly attuned to the need for water to ensure healthy ecosystems and avoid environmental degradation, and conservation groups are spearheading this agenda. To ensure an equitable sharing of water resources consistent with existing rights, a philosophy of collaboration and cooperation underlie the WaterSMART programs.

The Southwestern Willow Flycatcher (above) and the Bull Trout (below) depend on a good supply of clean water to survive and flourish.

Reliable and predictable deliveries of irrigation water are essential to farmers and ranchers in the West.

Two collaborative institutions – the Climate Science Centers and the Landscape Conservation Cooperatives – were established to foster greater collaboration. The accomplishments of these collaborative institutions are described in the chapter "A Comprehensive Landscape-Level Planning Strategy." The USGS and Reclamation are also partner agencies in the National Integrated Drought Information System, along with the National Atmospheric and Oceanic Administration, the Natural Resources Conservation Service, and a variety of State and local governments. There is a logical synergy in this partnership that provides leadership and an integrated approach to developing and using information on drought history, current conditions, and future projections to more effectively prepare for and manage drought.

The USGS approach to water use science, a key aspect of WaterSMART work, is also a highly collaborative process. The USGS works with water resource agencies in all fifty states, along with the U.S. Departments of Energy and Agriculture, the U.S. Census Bureau, and the U.S. Environmental Protection Agency, to name a few, in the process of estimating water use for the Nation. Finally, the USGS has commissioned a Water Census Advisory Committee under the auspices of the Advisory Committee on Water Information to provide guidance to the USGS on its WaterSMART activities through its twenty-seven member organizations and agencies. The Advisory Committee on Water Information also recently formed the Water Resources Adaptation to Climate Change Workgroup in response to recommendations from the Interagency Climate Change Adaptation Task Force's *National Action Plan: Priorities for Managing Freshwater Resources in a Changing Climate* and a report to Congress by the National Science and Technology

Council, *Strengthening the Scientific Understanding of Climate Change Impacts on Freshwater Resources of the United States*. The workgroup will promote, support, and provide information for effective management and protection of water resources in the United States as the climate changes and will advise Interior and other Federal agencies on water policy as it relates to climate change adaptation.

WaterSMART: An Integrated Approach

Over the last three years, the WaterSMART Program has enabled Interior and its partners to act in response to near-term and immediate water shortages, and also to plan for longer-term needs. This progress report describes the steps taken by Interior to implement a comprehensive program that supports balanced water management. The chapters that follow explain how the programs fit together to meet the water management goals Congress established through the SECURE Water Act and that Secretary Salazar outlined in his sustainable water management policy.

The first three chapters highlight Interior's immediate and near-term strategies to work with an array of partners to cost-effectively develop new water sources and make the most of existing supplies. Beginning with projects funded through the Title XVI Water Reclamation and Reuse Program, the report describes the ways wastewater is being recycled to provide supplemental water for a variety of important uses ranging from environmental restoration to municipal, irrigation, and industrial use. Next, examples of on-the-ground conservation and efficiency projects funded through WaterSMART Grants illustrate how communities stretch available water supplies to relieve pressure on existing sources. Interior's efforts to reduce its own water footprint are described as well. The combined results of Title XVI projects, WaterSMART Grants, and other conservation programs are then quantified to show the progress made toward Interior's goal of 730,000 acre-feet of water conserved annually by the end of 2013. Resources made available in 2010 and 2011 have facilitated an estimated 488,000 acre-feet in annual water savings through these critical water conservation and reuse projects.

The remaining chapters focus on longer-term planning and assessment efforts, collaborative institutions, and making the connection between energy and water. This section of the report starts with a description of the components of the WaterSMART Basin Study Program. The chapter spotlights just a few of the in-depth, watershed-focused investigations that exemplify how the program contributes to our understanding of water supply and demand and the potential effects of climate change on both. The next chapter examines the inextricable link between energy and water. As illustrated in several examples, Interior's project partners expect to achieve significant energy savings through water conservation and are also integrating renewable energy into existing water distribution facilities. The final chapter returns to the theme of longer-term planning. It focuses on the components of the new National Water Census, which will fulfill the congressional directive to conduct a National Water Availability and Use Assessment. This comprehensive review of the Nation's water resources will provide water managers and the public critical information on the quantity and quality of water, the many ways we use those resources, and the links between water and the health of the environment.

Sustainable water management takes into account water for recreation and for generating electricity.

Through WaterSMART, Interior and its partners are building on the successes of many existing activities. New complementary programs and funding mechanisms are also incorporated into WaterSMART to develop the science, technology, funding, and collaboration tools needed for us to better understand how much water we have, how much we need, and how to close the gap between the two. By continuing to expand on our efforts to date, Interior and its partners will make progress on the path toward a sustainable water future.

WATER REUSE: DEVELOPING AND SUPPLEMENTING LIMITED SUPPLIES

Throughout the country, wastewater effluent from households and commercial and industrial facilities is collected at wastewater treatment plants where it is treated prior to disposal. Most of those wastewater treatment plants accomplish "primary" treatment (removal of suspended solids and organic matter) and "secondary" treatment (further biological treatment and disinfection) before discharging treated wastewater into streams or rivers, or into the ocean.

Water management agencies can stretch existing drinking water supplies and help to ensure that growing water demands can be met through advanced treatment of wastewater. In particular, in communities where downstream users are not already relying on that treated wastewater as part of their water source, reuse of wastewater can become an additional supply, as well as a way to reduce discharges into particularly sensitive ecosystems. In its 2012 report, *Water Reuse: Potential for Expanding the Nation's Water Supply*, the National Research Council of the National Academies calculated that if all coastal discharges of municipal wastewater were to be reused, the additional water available would be equivalent to about 6 percent of the Nation's current water use.

> "Municipal wastewater reuse offers the potential to significantly increase the nation's total available water resources."
>
> —National Research Council of the National Academies (2012)

The State of California has made water reuse a fundamental part of its strategy to meet current and future demands and estimates that 900,000 to 1.4 million acre-feet of "new water" could be added to the State's supply by reusing municipal wastewater that currently flows to the ocean or saline bays. That potential new supply is about the amount of water needed for household use in San Diego and the surrounding metropolitan area. Similarly, the State of Texas is planning that its 2060 water supply will include about one million acre-feet of recycled water, meeting about 10 percent of projected demands.

> "If we didn't do reuse or conservation, we would have to go out and find more water. And the cost continues to go up to develop new water sources."
> —Michael Thane Director of Utilities City of Round Rock, Texas

In 1992, to implement Title XVI of Public Law 102-575, Reclamation established a program to address reuse of municipal wastewater, agricultural wastewater, and naturally impaired groundwater or surface waters. Through the Title XVI Water Reclamation and Reuse Program, Reclamation works with water management agencies in the western United States and Hawaii as they plan, design, and construct water reuse projects. Title XVI is an important part of WaterSMART. Projects completed through the program can improve efficiency, providing flexibility and diversifying the water supply. Reuse is often a drought-resistant supply, since sources such as treated municipal wastewater continue to be available during periods of water shortages. Title XVI projects provide growing communities with new sources of clean water while promoting water and energy efficiency and environmental stewardship. Reclaimed water can be used for a variety of purposes, such as environmental restoration, fish and wildlife, groundwater recharge, municipal, domestic, industrial, agricultural, power generation, or recreation.

Under the law, Reclamation may provide up to 50 percent of the cost of planning activities conducted by project sponsors. Once Congress authorizes Reclamation to participate in the construction of a particular project, Reclamation may provide funding, as available, for up to 25 percent of the total project costs, with funding generally limited to no more than $20 million per project. To date, 53 projects have been specifically authorized under Title XVI and two other projects have been undertaken through Reclamation's general authority to participate in research and demonstration projects.

Purple pipe prior to installation by the Napa Sanitation District in California. This pipeline, which is being constructed as part of Reclamation's Title XVI Program, will deliver recycled water for landscape irrigation, reducing reliance on local and imported surface and groundwater supplies and reducing the amount of effluent released to San Pablo Bay.

On-the-Ground Progress

In 2009, Reclamation awarded about $133 million in funding under the American Recovery and Reinvestment Act (ARRA) to Title XVI projects using a competitive selection process. As the Title XVI Program was incorporated into WaterSMART in 2010, Reclamation built on the process used under ARRA. New criteria were developed to prioritize congressionally authorized projects for the limited Federal funding available. Criteria place an emphasis on cost-effective projects that demonstrate they will reduce demands on existing facilities, result in a drought-proof supply, improve water quality or enhance habitat, incorporate renewable energy and energy efficiency, and meet other important program goals.

In 2011, Reclamation incorporated those criteria into new funding opportunities and invited sponsors of congressionally authorized projects to submit applications for funding. In 2012, based on project sponsors' feedback, Reclamation made revisions to streamline the process and again invited applications for funding.

Construction of an advanced water treatment facility, part of the $103 million water reuse project being built by the City of Oxnard, California with $20 million in Title XVI funding. The project, which is scheduled to be completed in 2013, is expected to result in 7,200 acre-feet of recycled water per year, through wastewater reuse.

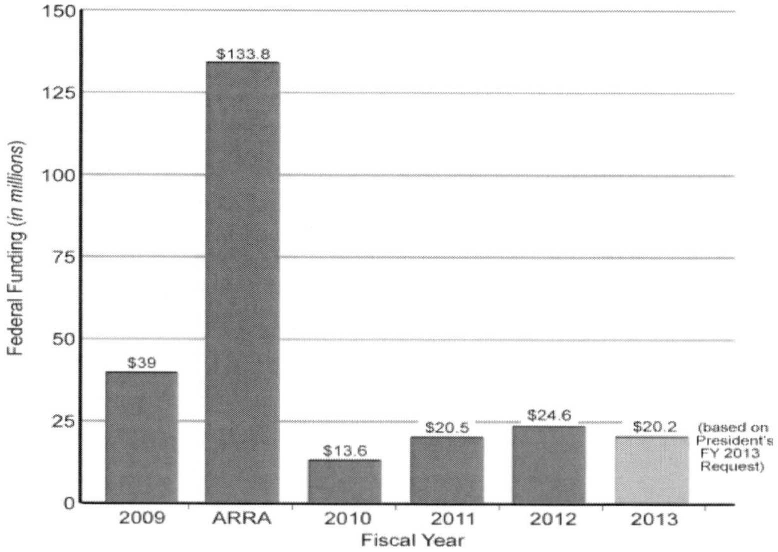

Title XVI funding since 2009. About $580 million in Federal funding has been provided for Title XVI through 2012, including $231 million since 2009. Taking into account non-Federal cost share provided by project sponsors and other entities, at least $2.3 billion in water reuse improvements have been constructed or will be finished by 2014.

Construction at the Santa Clara Valley Water District in California.

The results of recent funding can be seen in the number of congressionally authorized Title XVI projects that are finishing construction and beginning to deliver water. The Santa Clara Valley Water District in California expects to complete construction of its recycled water treatment facility in 2013. The new facility, built using $8.25 million in ARRA funding and more than $40 million in non-Federal funding, will treat secondary effluent to provide 9,000 acre-feet of recycled water per year to agricultural, industrial and landscape users to reduce dependence on existing potable water supplies and to reduce wastewater discharges.

As new projects become operational, the total amount of water delivered through Title XVI is increasing. Each year, project sponsors report the amount of water that their projects have delivered that year. Every acre-foot of recycled water delivered means that an equivalent amount of drinking water is available for other uses. Reclamation estimates that deliveries of recycled water each year will increase by almost 60,000 acre-feet from 2011 to 2013.

In New Mexico, for example, the **Albuquerque Bernalillo County Water Utility Authority** has now completed most of the water reuse construction activities that make up its Title XVI water reuse project. "Reuse and reclamation projects that allow for use of existing impaired water supplies are critical to managing water resources in the arid Southwest," says John Stomp, the Authority's Chief Operations Officer. "These supplies, like wastewater, used to be considered disposal problems, but they are vital resources if cleaned and used properly." Stomp points out that water reuse projects extend the life of other resources, reduce infrastructure cost for potable water supplies, and reduce the need for additional storage reservoirs.

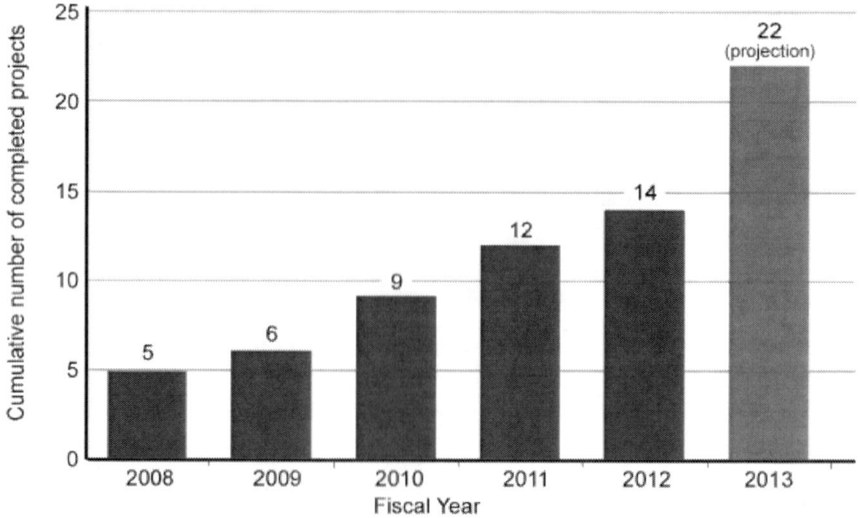

Completed Title XVI projects. Eight projects have finished construction since 2009, and eight others are expected to be completed in FY 2013.

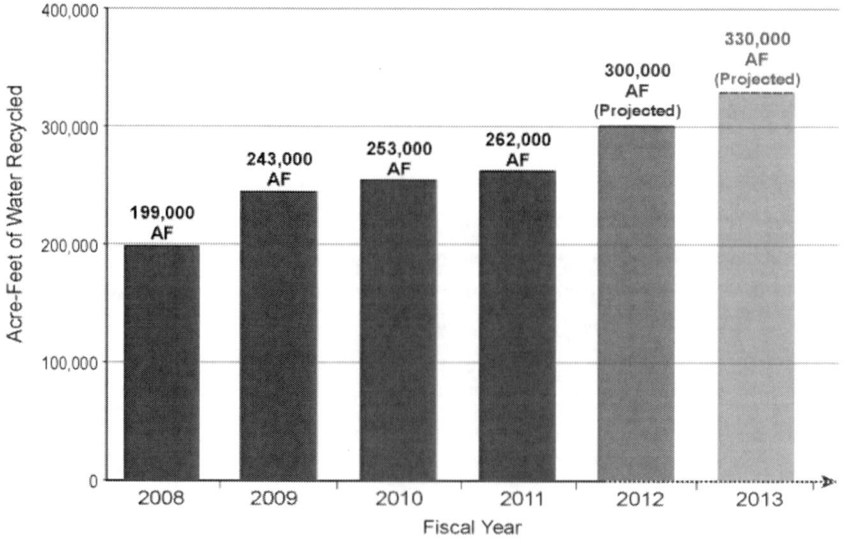

Recycled water. Water deliveries through completed Title XVI projects are expected to increase over the next few years as additional project phases come on line. In 2011, Title XVI project sponsors reported that they delivered about 262,000 acre-feet of water that year. Congressionally authorized Title XVI projects together have a maximum estimated capacity of about 735,000 acre-feet per year at build out.

"These supplies reduce the impact from water rights transfers from agriculture to municipal and industrial supplies and can free up water that could be used to support instream flows for endangered species."

—John Stomp Chief Operating Officer Albuquerque Bernalillo County Water Utility Authority

The Authority's Title XVI project includes several elements, including reuse of industrial effluent from a local microchip manufacturer for irrigation use, blending of San Juan-Chama surface water from the Rio Grande with industrial effluent for irrigation and industrial use, and filtration of treated municipal wastewater and delivery of recycled water to irrigation and industrial sites in Albuquerque. The project also includes a demonstration effort to remove naturally occurring arsenic from two existing water supply wells so that they could be brought back into production. "I think Title XVI has been one of the most successful endeavors we have undertaken in Albuquerque," says Stomp. "It has been a tremendously valuable tool."

In Texas, the **City of Round Rock** is planning to meet 10 percent of its projected 2050 demands through reuse and conservation. The City is building new facilities to perform additional treatment of municipal wastewater so that recycled water can be delivered to parks and schools for irrigation. To keep the cost of new distribution pipelines to a minimum, the City has identified recycled water customers located near its existing wastewater treatment plant, the source of the water to be reused. Michael Thane, the City's Director of Utilities and Environmental Services, points out that the project will allow the City to stretch drinking water supplies further, postponing the need for new surface water rights acquisitions and more expensive capital improvement projects.

The pump station at the Albuquerque Bernalillo County Water Utilities Authority's water reuse treatment facility, part of a Title XVI project.

About $950,000 was awarded for the City's Title XVI project this year, following $2 million in Federal funding awarded over the past few years for the $12.8 million project. "The Federal funding at twenty-five percent of the project cost is helping us in our entire water system capital improvement program," says Thane, "and is allowing us to build the system faster and sooner than anticipated." The project will begin to deliver recycled water in 2013 and is expected to provide 13,400 acre-feet annually once completed.

In southern California, the **Orange County Water District** completed construction of an indirect potable reuse project, the largest of its kind in the world, in 2008. The project now produces about 72,000 acre-feet of recycled water each year. The District brings treated wastewater to its Ground Water Replenishment System, where it performs a three-step advanced water treatment process – microfiltration, reverse osmosis, and ultraviolet light with hydrogen peroxide. The project is large – approximately $480 million in total costs, including $20 million in funding through Title XVI. "We took wastewater that would otherwise be discharged into the ocean and reclaimed it," says Mike Markus, the District's general manager. "Federal funding helps leverage dollars and helps push the project forward."

Distribution pumps (top) and construction of a recycled water pipeline (bottom) at the City of Round Rock's water reuse facility, which has been funded through the Title XVI Program.

The reverse osmosis building at the Orange County Water District's Groundwater Replenishment System, which was completed under the Title XVI program.

> "In our case, early on, we were able to say that we had authorization for $20 million in Federal funding and state bonds. Without that, we may not have been able to build the project."
>
> —Mike Markus General Manager Orange County Water District

The District manages the large groundwater basin that lies under northern and central Orange County, a source that provides water for more than 2.3 million Orange County residents. Recycled water helps to replenish the basin, which is otherwise recharged through imported water, the variable flows of the Santa Ana River, and local rainfall. "The only reliable supply to recharge groundwater is the Ground Water Replenishment System," says Markus. Water produced through the project is also used as part of a seawater barrier to prevent saltwater intrusion into the groundwater basin.

In addition to the Title XVI Program, Reclamation also makes funding available for smaller **WaterSMART Grants for Advanced Water Treatment**. Project sponsors such as States, Tribes, and irrigation districts apply for up to 50 percent of the cost of pilot or demonstration projects. Through WaterSMART Grants for Advanced Water Treatment, Reclamation helps to leverage funding that will accelerate the adoption and use of advanced water treatment technologies to increase water supply and to provide for long-term assessments of the potential for use of advanced water treatment. In 2010 and 2011 together, about $3.3 million was provided for seven new pilot and demonstration projects. Among the projects selected for funding in FY 2011, Loving County, Texas, is working to examine the use of wind-powered vapor compression to desalinate brackish groundwater, potentially providing a local, sustainable water source for the County. Four projects are expected to be completed by the end of 2012. No new WaterSMART Grants for Advanced Water Treatment were awarded in 2012.

Next Steps

More than $560 million in outstanding authorized Federal cost-share remains for authorized Title XVI projects. In 2011, Reclamation surveyed project sponsors about their plans and determined that sponsors of about 25 authorized projects will seek additional funding over the next two years. Together, those projects have an outstanding Federal cost-share of approximately $375 million. In FY 2013, Reclamation plans to conduct an update of that survey.

Construction of a 100 acre-foot capacity recycled water reservoir by the Sonoma Valley County Sanitation District, part of a Title XVI project to reduce reliance on groundwater supplies.
Below, Bureau of Reclamation Commissioner Michael L. Connor discusses water reuse with the District at the construction site.

As construction continues, Title XVI projects have the potential to make significant additional contributions to water supply sustainability. For example, the 36 Title XVI projects located in California together have the capacity to recycle more than 640,000 acre-feet of water per year – about 400,000 acre-feet per year beyond current production. Approximately $323 million in outstanding Federal cost-share remains for those currently authorized projects.

Interest in funding to study potential new water reuse projects is strong. In 2011 and 2012, Reclamation made funding available for the development of feasibility studies to assess new projects. Each of those years, more than 25 applications for funding were received – or about three times the number of studies that could be funded. Seventeen potential water reuse projects are currently being studied as a result of $2 million in Federal funding.

The President's budget for FY 2013 includes $20.2 million for the Title XVI Program. Again this fall, Reclamation plans to make funding available, subject to final appropriations, for the continued construction of congressionally authorized Title XVI projects.

WATER CONSERVATION AND EFFICIENCY: STRETCHING EXISTING SUPPLIES

Water conservation and efficiency are crucial to most western States' plans to ensure that water is available to meet demands into the future. The State of Texas plans to work toward increasing conservation and efficiency by about 1.5 million acre-feet per year by 2060, a volume that represents roughly 23 percent of the State's strategy to meet expected future demands. In California, the State has set a goal of a 20 percent reduction in urban water use by 2020 and is currently finalizing regulations to implement goals and best practices for agricultural water conservation. California has determined that higher levels of water delivery system performance can be achieved through infrastructure improvements such as regulating reservoirs, canal lining, additional system automation, and spill prevention. The State of Utah, similarly, has identified water conservation as a critical step in postponing development of new water supplies, and has set a goal of reducing per-capita consumption of water 25 percent by 2050 – or about 500,000 acre-feet of water per year in Utah alone.

Funding for non- Federal partners' water conservation and efficiency improvements is a fundamental aspect of WaterSMART.

Consistent with efforts being carried out at the State level, Reclamation's water conservation activities are a key aspect of efforts through WaterSMART to achieve sustainable water management. Through WaterSMART Grants – specifically Water and Energy Efficiency Grants, the main category for funding of on-the-ground projects – Reclamation provides 50/50 cost-shared funding, on a competitive basis, to non-Federal partners that wish to implement water conservation and efficiency projects. Irrigation and water districts, Tribes, States, and others with water or power delivery authority apply for funding of projects that can be completed in two to three years. Improvements include lining or piping of canals, installation of advanced measuring devices, irrigation system automation, installation of residential water meters, and activities that reduce urban water use, among other types of projects. Many WaterSMART Grant projects accomplish important program goals beyond water efficiency, including increasing the use of renewable energy, protecting endangered species, or facilitating water markets.

Construction by the Davis and Weber Counties Canal Company in Utah to line a canal with reinforced concrete. As a result of the improvements, about 7,100 acre-feet of water that was previously lost to seepage can be used to alleviate water shortages in the area. The $7.3 million project, which was constructed using $3.6 million in ARRA funding, is now complete.

On-the-Ground Progress

In 2009, more than $44 million, including $40 million in ARRA funding for projects that have now been completed, was provided to non-Federal project sponsors for water conservation and efficiency improvements. The following year, Reclamation's existing grant program was expanded as it was incorporated into WaterSMART. In 2010 and 2011 together, Reclamation awarded more than $38 million in WaterSMART Grants to project sponsors to implement 95 new on-the-ground water management improvements. Once the sponsors of those projects contribute non-Federal funding to each of those efforts, more than $138 million in improvements will result. To date, about 25 of those projects have been completed, with the others under construction or will begin construction soon. In addition, 32 new projects have been selected for funding in 2012. Reclamation has awarded a total of $11.1 million for those new projects, with construction of additional improvements to begin this winter and spring.

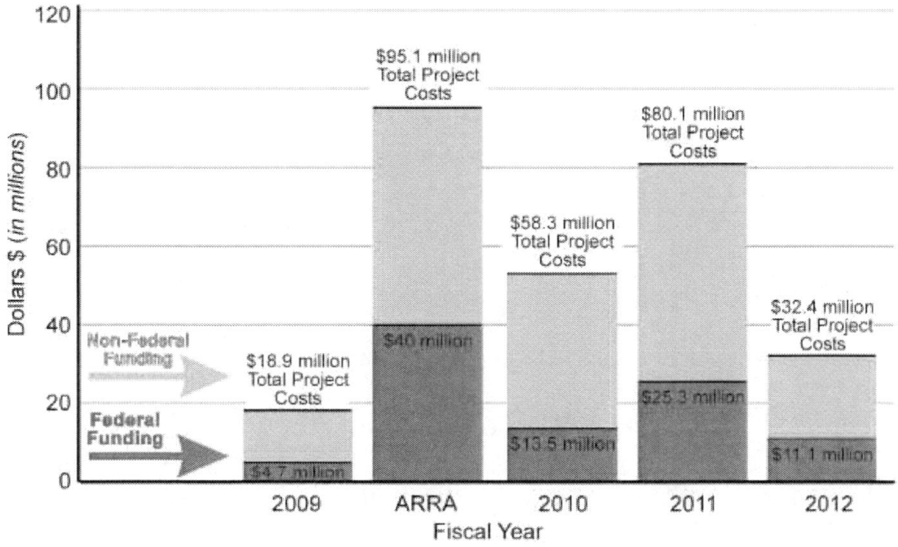

Leveraging Funding. Through WaterSMART Grants, Federal funding (shown here in dark green), is leveraged with at least 50 percent non-Federal funding (light green) to accomplish on-the-ground water management improvements.

Automated gates like the one shown (left) here during installation, along with a radio-based control system, will allow the Authority to operate a portion of its system in real time, avoiding spills and overdeliveries. This gate and others were installed as part of a WaterSMART Grant.

> "With Federal funding, we were able to add SCADA and automated spill site monitoring to the project, meaning that we will be able to get a lot closer to being able to verify the savings of the new gates."
>
> —John McLeod Project Manager Lower Colorado River Authority

Sponsors match Federal funding dollar for dollar on all WaterSMART Grant projects, and many sponsors are able to secure non-Federal funding beyond that required 50 percent level. WaterSMART Grants therefore work to leverage funding: since 2009, about $94 million in Federal funding has been used to implement more than $280 million in water management improvements across the West.

In Texas, the **Lower Colorado River Authority** has made a significant effort over the last few years to improve the efficiency of the 1930s era water delivery system in its Gulf Coast Irrigation Division. The need for action has been highlighted by recent drought conditions: for the first time in 100 years, water has not been available to farmers served by the system. The Authority recently completed a long-term strategic plan that identified steps necessary improve the reliability of water for farmers. The plan included recommendations for a new Supervisory Control and Data Acquisition (SCADA) system and automation of major check structures within the irrigation canal system.

An existing canal gate (above) at the Lower Colorado River Authority's Gulf Coast Irrigation Division in Texas. Operation of manual gates such as this one requires that an employee travel throughout the 300-mile canal system, limiting the Authority's capability to deliver precisely the amount of water needed by farmers.

In 2010, the Authority began implementation of the plan through a $257,000 WaterSMART Grant to complete a $535,000 improvement project. The Authority has automated eleven new canal gate structures and installed a radio-based communication system to remotely operate those gates. The project also included a new SCADA system that automatically controls water levels in the canal system and collects and displays real- time canal data for system operations. With the new system in place, the Authority is able to maintain precise control of water levels in the canals for accurate water delivery to each farmer's field.

The Authority's project manager, John McLeod, points out the cost-effectiveness of the project relative to other options. "We identified this as one of the lowest cost improvements that could be made to save water," says McLeod. The project has now been completed, and is expected to result in 2,560 acre-feet of water savings annually. Conserved water will be marketed to existing municipal water customers through an agricultural- to-urban water transfer program established under 1999 Texas law. The Authority also estimates that the project will reduce energy consumption by approximately 132,000 kilowatt hours per year, primarily through reductions in pumping.

In Oregon, the **Three Sisters Irrigation District** was selected for an $859,000 WaterSMART Grant in 2011. That funding is being used to complete work on a $3.3 million phase of the District's ongoing collaborative effort with the Deschutes River Conservancy to improve its delivery system. Over the last century, irrigation demands have been much greater than the water available from Whychus Creek, and parts of the creek that serve as critical habitat for steelhead and Chinook salmon often dried up in August or September. The District estimates that as much as 50 percent of the water diverted for irrigation through open, unlined canals was seeping into the porous, volcanic soils before it reached farms, presenting a clear opportunity for improvements. "Our creek happens to be a major spawning territory for summer steelhead," says Marc Thalacker, the District's general manager, "and there has been an incredible amount of attention paid to Whychus Creek."

The unlined Uncle John diversion prior to work by the Three Sisters Irrigation District to install buried pipe. This open, unlined canal was used to transport water from Whychus Creek, in the Deschutes River Basin in Oregon, for irrigation.

Through piping projects such as this one, the Three Sisters Irrigation District and the Deschutes River Conservancy have been able to protect more than 20 cubic feet per second of Whychus Creek flows, effectively meeting the Oregon Department of Fish and Wildlife's instream flow target for steelhead and Chinook salmon.

The WaterSMART Grant project includes work to convert almost four miles of those canals to buried pipe, which is expected to result in 750 acre-feet of water savings annually. That conserved water will be marketed for a protected instream right to restore Whychus Creek streamflows. Restoring fish passage and re-introducing salmon and steelhead to the region, in turn, was a key part of the agreement to relicense the Pelton-Round Butte Hydroelectric Project in the Deschutes Basin. In 2011, the Deschutes River Conservancy surpassed the Oregon Department of Fish and Wildlife's instream flow target for the upper reach of Whychus Creek–20 cubic feet per second–through the cooperative efforts of landowners and the District, including work completed as part of previous WaterSMART Grant projects. Water savings resulting from the District's current project will add to those instream flows.

> "Through projects like these, we are creating sustainable agriculture, and also addressing Clean Water Act and Endangered Species Act concerns. We are also avoiding future conflict, and we are getting there because we are able to restore the stream."
>
> —Marc Thalacker General Manager Three Sisters Irrigation District

As part of the project, the District will also install a 950-kilowatt capacity turbine generator, a renewable source of energy that will supply 3.1 million kilowatt-hours of electricity each year. Revenue from generation of electricity is also a potential source of funding for future water conservation improvements, according to Thalacker. "These projects would not be happening without this Federal funding. It would be an incredibly slow process. Instead, we're almost at the finish line." Construction of the project is expected to be finished in 2013.

Collaborative Projects for the Bay-Delta

In 2011, Reclamation and the Department of Agriculture's Natural Resources Conservation Service began a new partnership to leverage funding for water delivery agencies and agricultural producers to California. The joint effort is an important part of meeting the 2009 Interim Federal Action Plan for the San Francisco Bay/Sacramento-San Joaquin Delta. Through a competitive process, Reclamation made $4.1 million in WaterSMART Grants available to five irrigation districts so that improvements that save water or improve water management could be made in the systems that deliver water to farmers.

NRCS, in turn, committed to provide $7 million to farmers who receive water from those districts so that on-farm conservation improvements could also be made throughout those five districts. Two of those projects have now been completed.

The **South San Joaquin Irrigation District** in California has combined $1 million in WaterSMART funding with non-Federal funding to complete over $12.9 million in improvements to its water delivery system. In the past, growers experienced difficulties converting from flood to pressurized irrigation because of moss and algae in the water delivered by the District via open channel. Some of those growers also relied on groundwater, which was becoming more salty and difficult to pump. Through the WaterSMART Grant project, the District built a state-of-the-art, pressurized irrigation system, replacing open channels and incorporating automated controls and updated metering technology. "It really was a collaborative process between the District and growers to come up with solutions," says Jeff Shields, the District's general manager. "We worked together from the beginning to design the system in a way that worked for everyone."

Installation of pipe at the South San Joaquin Irrigation District, part of a $12.9 million project to convert the District's system from open channel delivery. With assistance from NRCS, farmers who receive water from the District are taking advantage of the new automated, pressurized system to convert from flood irrigation to more efficient sprinkler and micro-drip systems.

Canal lining by the Firebaugh Canal Water District in California. The project, which is now complete, is intended to prevent seepage losses and to control sediment, setting the stage for growers to work with NRCS to install high-efficiency drip irrigation systems.

> "We had a viable project and a vision for how to do something innovative to solve a groundwater problem, to make the system more energy efficient for the district and for farmers, and to conserve water. But the project needed the economic support from Reclamation to bridge the shortfall in the budget set by the Board."
>
> —Jeff Shields General Manager South San Joaquin Irrigation District

The improvements allow for precise measurement and accounting of water use, and also help the District to capture additional agricultural run-off for irrigation re-use. About 10 percent of the 50,000 acres of land within the District have now worked with NRCS to convert from flood irrigation to sprinkler or micro-drip irrigation systems that result in significantly less on-farm water use, with additional on-farm work continuing. "I am extremely happy with this new micro-irrigation system," says Samuel McGinnis, one of the landowners working with NRCS. "Now that I have converted from flood, I couldn't imagine going back."

Through a $1 million WaterSMART Grant that was also part of the joint Reclamation-NRCS effort in California, the Firebaugh Canal Water District has completed a $2.8 million project to address seepage losses and sediment in the water it delivers to growers. By lining 2.6 miles of earthen canal with concrete, the District was able to decrease suspended silts so that growers can reduce the back-flushing and filtering needed for efficient on-farm irrigation systems. Growers are currently working with NRCS to implement drip and sub-surface drip irrigation systems. Jeff Bryant, the District's general manager, points out that the project is also an important part of the District's efforts to meet sediment load targets. The District is a participant in the Grassland Bypass Project, through which Reclamation and the San Luis & Delta-Mendota Water Authority work to prevent discharge of subsurface agricultural drainage water into wildlife refuges and wetlands in central California.

> "If we weren't able to complete projects like this one, we would have to find other ways to manage subsurface drain flows, such as the need to purchase additional land to expand our reuse area. Canal lining gives us multiple benefits; we can address subsurface draining issues and we are also able to generate conserved water that's transferable to South of Delta contractors to mitigate chronic water shortages."
>
> —Jeff Bryant General Manager Firebaugh Canal Water District

In May 2012, Reclamation selected a second round of projects to be funded jointly with NRCS as part of the Bay-Delta Restoration Program. Ed Burton, the NRCS State Conservationist for California until his retirement earlier this year, points out the value of the partnership for both agencies. "Reclamation has done excellent work helping to install new and efficient water delivery systems in rural communities but we had yet to find a way to marry up this work with what NRCS does helping farmers conserve on-farm water," says Burton. "These projects have helped us accomplish just that. Farmers have been able to install new and better irrigation systems that are linked to new Reclamation water-delivery infrastructure and together we have helped rural communities be better stewards of limited and precious water supplies."

Beginning this year, Reclamation and NRCS are also building on their experience in California to coordinate on water conservation projects west-wide. In 2012, funding criteria for Reclamation's WaterSMART Water and Energy Efficiency Grants were expanded to identify proposals for water delivery improvements that could also be expected to lead to on-farm improvements. Four projects selected for funding will enable farmers to make additional on-farm improvements in the future. Reclamation and NRCS are currently working to assess each of those new projects to identify opportunities for NRCS involvement.

The Water Conservation Field Services Program

Smaller-scale efficiency improvements (up to $100,000 in Federal funding) and water conservation planning are funded through the Water Conservation Field Services Program, which was established in 1996 and was incorporated into WaterSMART in 2011. In 2010 and 2011, Reclamation provided 50/50 cost-shared funding for 129 projects through the program. Sixty-nine of those projects are now complete. In Utah, the City of St. George is using $25,000 in Federal funding to carry out a $50,000 program to provide rebates to customers who install pressure regulating valves on their irrigation systems. In Arizona, Bullhead City is using $25,000 to revise its five-year water conservation plan, a key activity carried out under the program. Additional projects have been selected for 2012 funding, which were awarded in September 2012.

System Optimization Reviews to Identify Future Improvements

Through System Optimization Reviews, Reclamation also makes 50/50 cost-shared funding available for entities that deliver water to assess the potential for water management improvements and identify specific ways to implement those improvements in the future. In 2010 and 2011 together, Reclamation provided $1.6 million to assist irrigation districts, municipalities, States, and Tribes in assessing the potential for future water management improvements. In the Lower Rio Grande Valley of Texas, the **Harlingen Irrigation District** operates 40 miles of open canal and 200 miles of pressurized lines to deliver water from the Rio Grande to the city of Harlingen and a large number of farmers in the area. The District has implemented a number of water efficiency measures in the past and wanted to determine which additional improvements would provide the best return on investment. Using $73,000 in Federal funding, the District completed a System Optimization Review that evaluated the current conveyance efficiency of its canals and the potential benefits of Supervisory Control and Data Acquisition, automation, control structures, river and ancillary pumping, off-channel storage, and re-regulation of storage. According to Tom McLemore, the District's project manager, conducting a System Optimization Review was an important step in determining which types of improvements would be worth the cost. "The information we gathered will allow us to make improvements to the district and have data to support it," says McLemore. "It also allowed us to put a dollar value on improvements to determine whether they are good investments."

> "The nice thing about the System Optimization Review is that it allowed us to step back, look at the district as a whole, and try to focus on where we could best spend funding on water efficiency improvements."
>
> —Tom McLemore Project Manager Harlingen Irrigation District

WaterSMART Clearinghouse

Interior has also established the web-based WaterSMART Clearinghouse, a tool for sharing water conservation information with the public and water resource managers. Through the Clearinghouse, information on best practices, case studies, education and outreach, sources of grants and other financial resources, research, and water data are available to the public. The 2010 Secretarial Order that established WaterSMART recognized that working toward sustainable water supplies involves coordination and collaboration – not only among Federal agencies but also State, Tribal, and local water resource managers – and called for creation of a public resource to coordinate and integrate strategies.

The WaterSMART Clearinghouse, which can be accessed online at: www.doi.gov/watersmart, is used to share best practices for water conservation and sustainable water strategies.

Web site users can search the Clearinghouse by topic or location, including State, river basin, or Tribal area to find information on subjects such as revolving loan programs, water treatment and engineering research, forecasting, and flow calculation tools. Water resource managers can submit information and website links to be added to the Clearinghouse. The initial phase of the Clearinghouse has been limited to gathering information from governmental agencies and land grant universities. Interior plans to look at expanding the types of entities that can submit information to the Clearinghouse in the future.

Reducing Water Use at Interior's Facilities

Interior is making considerable strides in conserving water use at its own facilities as well. The Secretarial Order that established WaterSMART requires that Interior meet and exceed the water consumption goals of Executive Order 13514, *Federal Leadership in Environmental, Energy, and Economic Performance*. Under the terms of the executive order, Interior must reduce its potable water use intensity 26 percent in terms of gallons per gross square foot by FY 2020 relative to a 2007 baseline. Interior must also reduce industrial, landscaping, and agricultural water use by 20 percent, measured in gallons, by FY 2020 relative to a 2010 baseline; promote and implement water reuse strategies; and achieve objectives established by the U.S. Environmental Protection Agency's Stormwater Management Guidance. Interior is making good progress towards these goals. Thanks to bureau efforts, Interior achieved an 11.2 percent reduction in potable water use intensity in FY 2011 exceeding the goal of 8 percent.

To meet Executive Order 13514, Interior is working to reduce its potable water use intensity 26 percent by 2020.

Interior's bureaus are pursuing a broad range of water savings strategies. Bureaus are implementing water efficiency best management practices recommended by the Department of Energy's Federal Energy Management Program and are using EPA's WaterSense products, including toilets, faucets, urinals, and showerheads. Bureaus are installing low-flow or ultra- low-flow plumbing fixtures in all new facilities. Landscaping design and construction is increasingly emphasizing native plant species, minimizing or eliminating artificial irrigation, and maximizing efficiency of necessary irrigation through the use of drip systems, precipitation detection systems, and optimal timing. Many Bureau facilities are putting in place water reuse strategies, such as treating gray water from restroom sinks, showers, or laundry, and then using that water for landscape irrigation. Finally, to comply with EPA's Stormwater Management Guidance, new buildings are designed to maintain or restore the predevelopment hydrology of the site with regard to temperature, rate, volume, and duration of flow using site planning, design, construction, and maintenance strategies.

Interior is in a unique position to demonstrate the importance of water conservation and the innovative water saving strategies that can be put to use. Interior owns and operates approximately 47,000 buildings, and over 70,000 structures, at 2,400 locations across the United States, Puerto Rico, and U.S. territories. About 477 million visits are made to these facilities each year, which gives Interior a tremendous opportunity to showcase water conservation efforts and influence behavior. Many bureaus showcase energy efficiency, renewable energy, and water conservation projects through kiosks and interactive displays at visitor centers. Public information related to drought and water conservation is also available at many Interior facilities. The Secretary's Youth Initiative further promotes conservation efforts and instills involvement with today's youth who will play a key role in water management for future generations.

Newly constructed or renovated visitor centers provide excellent examples of some of the proactive water saving strategies employed by Interior. The U.S. Fish and Wildlife Service recently opened a new visitor center at the **Neosho National Fish Hatchery** in Missouri. In addition to including many energy saving features, the Neosho building uses low-flow plumbing fixtures that conserve 28,225 gallons of water annually. Also, using hatchery spring water for large aquariums eliminates the need to de-chlorinate potable city water, while water-efficient landscaping with native plants and forbs avoids chemical use and irrigation. Stormwater containment and drainage swales conserve even more water and the implementation of a stormwater management plan reduces impervious cover, promotes infiltration, and captures and treats the stormwater runoff from 90 percent of the average annual rainfall using acceptable best management practices. Finally, an abandoned fish pond was used as a stormwater collection/retention pond for runoff by re-grading and planting aquatic vegetation.

In 2011, the National Park Service began the restoration of the original water cisterns at Castillo de San Cristobal and Castillo San Felipe del Morro at **San Juan National Historic Site, Puerto Rico**. Harvested rain water will be directed out of the cisterns to a treatment system that will allow water to be used for toilets, showers, and hose bibs. When completed, the cisterns will have a storage capacity of more than 1.2 million gallons and will save $12,000 annually in water costs.

Another example of water conservation comes from Reclamation's **Snake River Area Office** in Boise, Idaho, which reduced landscape water usage by replacing numerous water head sprinklers with drip irrigation and increasing mulching to retain moisture. The existing wood bark was replaced with two inches of permabark to increase moisture retention. Composting was established on site with grass clippings and leaves, which will then be put back into flower beds and around trees and bushes for added water retention. An outside water meter has been installed to measure and record irrigation water use separately from potable building water. This device will document water use and encourage reductions in potable water used for landscaping.

The Neosho National Fish Hatchery in Missouri, where innovative water conservation features have been incorporated into the design of a new visitor center.

Cisterns at the San Juan National Historic Site in Puerto Rico. The original cisterns have been restored and are part of a system used to harvest and treat rainwater for use on site, saving water costs at the facility.

The National Park Service has nearly completed the first of three phases to restore turf and improve drainage on the **National Mall in Washington, D.C.** The first phase includes gutters surrounding three Mall Lawn Panels and two large underground cisterns that will collect stormwater. Meanwhile, a below-grade pump station and a new irrigation system will distribute the collected water. When all three phases of the project are completed, the amount of potable water required for irrigating the Mall lawn panels will be reduced by approximately 7.56 million gallons per year (a savings of 67.5 percent). The collection of stormwater will also improve regional water quality by reducing the amount of flow from the Mall into the Washington area combined sewer system and keep raw sewage and pollutants out of Chesapeake Bay tributaries.

These innovative projects demonstrate Interior's ongoing commitment to conserving water at Department facilities.

Construction of turf and drainage improvements on the National Mall in Washington, D.C.

Next Steps to Address Water Conservation and Efficiency

WaterSMART Grant projects selected for award this year will begin construction this winter and spring. The President's budget for FY 2013 includes a request of $21.5 million for WaterSMART Grants. Reclamation plans to announce a new funding opportunity this fall so that potential new projects can be selected as soon as funding is made available.

Interest in WaterSMART Grant funding continues to be much greater than available funding. Over the last three years, only a portion of highly ranked proposals could be selected for funding. In 2012, for example, 32 of the 167 proposals submitted by project sponsors could be selected for funding.

Even as interest in WaterSMART Grant funding remains high, Reclamation is working to ensure that opportunities are as inclusive as possible and that grant categories align with interest among applicants. For that reason additional outreach efforts are planned for 2013 to discuss the program with potential project sponsors.

This year, Reclamation is also working to record the key internal requirements and procedures used to select, award, and administer WaterSMART Grants. In 2012, Reclamation developed temporary Directives and Standards so that requirements are documented as soon as possible to ensure consistency and transparency. In 2013, as Reclamation takes steps to incorporate permanent Directives and Standards for the program into the Reclamation Manual, the public will be asked to review draft documents and to provide comments.

Interior is also moving forward with implementation of the Cooperative Watershed Management Program, which has been established to support the formation and development of locally led watershed groups and to facilitate the development of multi-stakeholder watershed management projects. The program is an Interior initiative that will involve participation by multiple agencies, including Reclamation, the USGS, and the Bureau of Land Management, and will involve strong partnerships with States. The purpose of the program is to improve water quality and ecological resilience, conserve water, and to reduce conflicts over water through collaborative conservation efforts in the management of local watersheds. The Cooperative Watershed Management Program contributes to the WaterSMART strategy by providing funding to watershed groups to encourage diverse stakeholders to form local solutions to address their water management needs.

The CWMP will be implemented in three phases. This year, funding was made available for establishment or expansion of a watershed group. Interest in this new program is strong. In July 2012, Reclamation received 30 applications for funding to establish or expand watershed groups, together representing a request for approximately $1.7 million. Eight awards totaling $333,500 in Federal funding were made at the end of September 2012. Opportunities for assistance in implementing watershed management projects are anticipated in 2013 or 2014, depending on available funding.

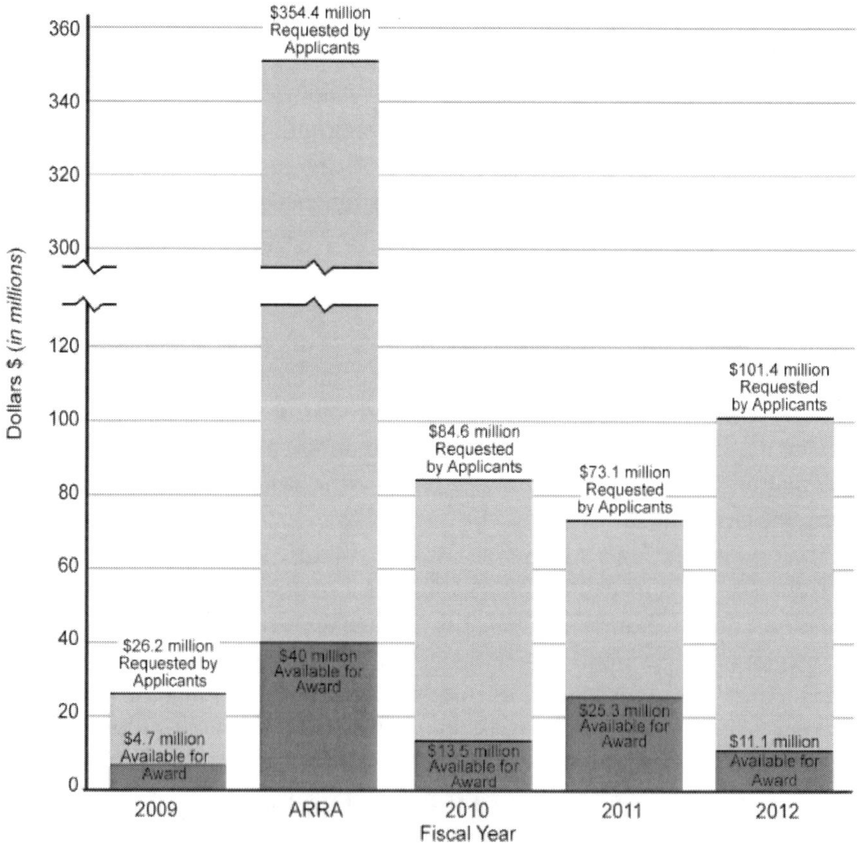

Continued Interest in WaterSMART Water and Energy Efficiency Grants. Over the last four years, requests have been far greater than the Federal funding available for award. In 2012, applicants together requested over $100 million, or about nine times the available funding.

The Priority Goal for Water Conservation: 730,000 Acre-Feet by the End of 2013

Beginning in 2010, Interior and other Federal agencies established a series of outcome-based, near-term performance goals to help focus on key initiatives. One of the five goals identified as Interior's highest priorities for 2010-2011 was a new Priority Goal for Water Conservation: To enable capability to increase the available water supply up to 350,000 acre-feet by the end of 2011 through Reclamation's conservation-related programs.

By September 30, 2013, Interior will enable the capability to increase the available water supply for agricultural, municipal, industrial, and environmental uses in the western United States through Reclamation's water conservation programs to 730,000 acre-feet, cumulatively since 2009.

For 2012 and 2013, Interior extended some of its goals, including the Priority Goal for Water Conservation. Interior and Reclamation are currently working to address that expanded Priority Goal: To enable the capability to increase the available water supply through Reclamation's water conservation programs to 730,000 acre-feet by the end of September 2013.

Contributing Activities

Initially, Reclamation addressed the Priority Goal for Water Conservation through two significant WaterSMART activities: WaterSMART Grants and the Title XVI Water Reclamation and Reuse Program. Both of those programs result in projects that have quantifiable water savings. Once other programs were reviewed, Reclamation concluded that the Water Conservation Field Services Program and several other regionally-focused water conservation efforts also result in quantifiable water savings that can contribute toward the goal.

In California, CALFED Water Conservation Projects provide state-wide benefits through water conservation, including quantifiable water savings that contribute to the Priority Goal. CALFED is a combined State of California and Federal program focused on restoring the Sacramento-San Joaquin Delta's fragile ecosystem and improving water supply reliability for urban and agricultural water users. Water use efficiency from districts linked to the Bay-Delta water supply can result in significant benefits to water quality, water supply reliability, and in stream flows.

Through the Yakima River Basin Water Enhancement Project in Washington, irrigation districts implement structural and nonstructural measures to increase the reliability of the irrigation water supply and enhance streamflows and fish passage in the Yakima River basin. Facility modifications and other steps through the project to reduce diversions lead to quantifiable water savings, which contribute to the Priority Goal.

The Sunnyside Division Board of Control's re-regulation reservoir, under construction (top) and after being filled (bottom). The reservoir was constructed as part of a project that contributes to Interior's Priority Goal for Water Conservation.

The multi-agency, multi-interest group Upper Colorado River Recovery Implementation Program includes the Orchard Mesa Irrigation District Canal Automation Project, which is intended to augment flows in a critical reach of the Colorado River in western Colorado. Through the project, a new canal automation system will allow for irrigation diversions that more closely match actual water demands, resulting in water savings that contribute toward the Priority Goal. This reach of the Colorado River, upstream from the confluence with the Gunnison River, is heavily impacted by multiple Federal and private irrigation diversions.

Water conserved through the project will be redirected to an existing hydroelectric power plant that discharges back to the river upstream of the depleted river reach, augmenting flows in critical habitat and increasing clean renewable energy. This large, multi-year project includes state and local cost-shared funding.

To address seepage losses in its open canal, the Fort Shaw Irrigation District in Montana is converting a section of that canal to buried pipe. The project is being completed through a 2011 WaterSMART Grant, and water savings resulting from the improvement contribute toward Interior's Priority Goal for Water Conservation.

Calculating Water Savings

Each year, Reclamation reports the water savings expected to result from water conservation activities funded that year. Water savings for each funded project are based on estimates of the number of acre-feet expected to be conserved each year once that project becomes operational. The estimates are provided by non-Federal project sponsors in applications for financial assistance, feasibility reports and other project planning reports, and National Environmental Policy Act documents, and are based on water measurement and accounting records, calculations by the project sponsor's engineering sources, hydrologic modeling, statistical analysis of historic climatic data, and other information.

For example, in January 2011, the Fort Shaw Irrigation District in Montana submitted an application for a new WaterSMART Grant. The District planned to convert a section of open, unlined canal to a new pipeline to address water losses due to seepage, and requested $103,000 for the $222,000 project.

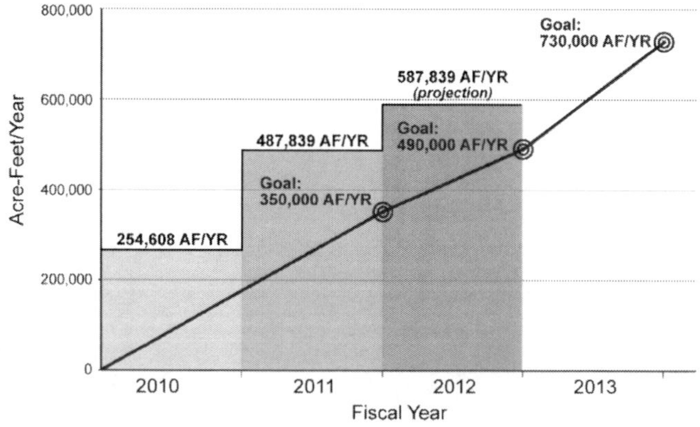

As part of its application materials submitted to Reclamation, the District made an estimate that the project would lead to savings of 4,158 acre-feet of water each year once constructed. A team of Reclamation employees with technical expertise then assessed the District's water savings estimate, assumptions, and supporting calculations as part of the competitive review of all applications for funding. The project was selected for funding in the spring of 2011. Once Reclamation and the District executed a financial assistance agreement for the project in September 2011, water savings of 4,158 acre-feet expected to result from the improvement were reported toward the FY 2011 Priority Goal target. The project is currently under construction.

Progress toward the Priority Goal

Projects funded in FY 2010 and FY 2011 are expected to result in 487,839 acre-feet of water savings per year, above the original Priority Goal of 350,000 acre-feet per year by the end of FY 2011. Reclamation anticipates that at least 100,000 acre-feet in additional water savings every year will result from FY 2012 projects. Based on water savings reported through FY 2011 as well as those projections for FY 2012, Reclamation and Interior are on track to achieve the goal of 730,000 acre-feet of water savings.

Steps to Increase the Accuracy of Water Savings Estimates

In 2011, Reclamation began an effort to confirm project sponsors' water savings estimates by conducting on-site tests for a small number of projects, on a pilot basis, prior to construction. One set of tests is being conducted for the Florida Farmers Ditch Company in Durango, Colorado, which has been awarded a WaterSMART Grant to line 1.8 miles of open canal with concrete. Before construction began, Reclamation conducted early and late season "inflow/outflow" tests at the site. Using acoustic Doppler Devices, Reclamation measured the flow in and out of the canal section on each occasion, as well as channel depth. By subtracting the amount of water flowing out of the canal from the amount that entered that section of canal, the seepage rate was calculated. Once the project has been completed in 2013, Reclamation will return to measure post-project seepage rates. Similar inflow/outflow tests will be completed before the East Hoytsville Ditch Company in Utah begins work to convert 15 miles of open canals to pipe to confirm the water savings resulting from that project.

Interior reported 254,000 acre-feet toward the goal in FY 2010 and a cumulative amount of 487,000 acre-feet at the end of FY 2011, exceeding the two-year goal of 350,000 acre-feet. Based on projections, that cumulative total will rise to over 587,000 acre-feet per year at the end of FY 2012, well above the target for that year.

Another type of testing – "ponding" tests – will be conducted before the East Columbia Basin Irrigation Company in Washington converts about 85,000 feet of open ditch to pipeline to addressing current seepage losses. By placing a temporary dam in the ditch, the drop in water elevation and the amount of water necessary to maintain a constant water level can be measured, leading to a calculation of the current rate of seepage.

Once the first round of tests has been completed in the fall of 2012, Reclamation will assess the results of this pilot effort. If necessary, additional adjustments to selection criteria and review processes will be considered to further improve water savings estimates.

Interior expects to achieve its total goal of 730,000 acre-feet per year – about the amount of water needed for household use in Denver, Colorado and the surrounding metropolitan area each year – by September 2013.

In 2012, Reclamation also took steps to increase consistency in the water savings estimates developed by WaterSMART Grants project sponsors. Reclamation now includes a series of specific questions for applicants that reflect the particular steps necessary to develop accurate water savings estimates according to the type of project proposed. For example, applicants proposing canal lining or piping projects to address seepage losses are asked to provide detailed descriptions of the ponding tests or inflow/outflow tests conducted by the applicant. Applicants proposing installation of municipal meters, on the other hand, are asked for detailed information on existing water use patterns, any studies conducted in the region on the effect of meter installation on water use, and any other assumptions made.

Next Steps

Once FY 2012 funding has been awarded under the programs that contribute toward the Priority Goal for Water Conservation, final water savings will be reported for that year. Reclamation will also build on the steps taken so far to conduct on-site, before-and-after tests to confirm water savings estimates for a second year. Reclamation is currently reviewing the WaterSMART Grant projects selected for FY 2012 funding and will work with project sponsors this fall to identify a small number of new projects to continue that pilot effort.

A COMPREHENSIVE LANDSCAPE–LEVEL PLANNING STRATEGY

Sustainable water management requires understanding and planning for changes in the future. Taking advantage of the program authority established by Congress through the SECURE Water Act, the Secretary of the Interior included a comprehensive landscape-level planning strategy as part of WaterSMART. The WaterSMART planning strategy consists of several programs designed to help Interior and its partners identify current and future water supply imbalances. Three primary activities described in this chapter form the WaterSMART Basin Study Program: West-Wide Climate Risk Assessments, Basin Studies, and Landscape Conservation Cooperatives. Climate Science Centers, which complement and inform the work of Landscape Conservation Cooperatives, are also described. Together, these activities allow Interior and its partners to integrate evaluations of water supply needs, the availability of water supplies, the anticipated risks to water availability, and the impacts of reduced availability and increased demands.

West-Wide Climate Risk Assessments

The Bureau of Reclamation conducts reconnaissance-level West-Wide Climate Risk Assessment studies to understand how changes in the climate might affect water supplies across the western United States. The West-Wide Climate Risk Assessment studies also provide information on how changes in the climate might affect the demand for water because the combined impact of these changes will lessen the reliability and predictability of water supplies. Reclamation needs the information developed through West-Wide Climate Risk Assessment studies to understand how variations in climatic conditions may change the management of river system operations. But non-Federal water managers benefit from having this information, too. For this reason, the studies provide baseline climate projections for use in WaterSMART Basin Studies and other water resources implementation studies.

The West-Wide Climate Risk Assessments have also produced valuable tools that provide immediate and direct benefits to water managers. Reclamation collaborated with Federal and non-Federal entities to develop two online computer applications water managers can use to plan for altered water supply conditions. The *Streamflow Projections for the Western United States*

Web site allows users to access projected streamflow data from nearly 200 locations along streams and rivers in the western United States. The streamflow projections provided through this application are based on the *Bias Corrected and Downscaled WCRP (World Climate Research Program) CMIP3 (Coupled Model Intercomparison Project Phase 3) Climate and Hydrology Projections* archive data, developed collaboratively by Reclamation; Lawrence Livermore National Laboratory; Climate Central, National Oceanic and Atmospheric Administration, National Weather Service, Colorado Basin River Forecast Center; Santa Clara University; and Scripps Institution of Oceanography.

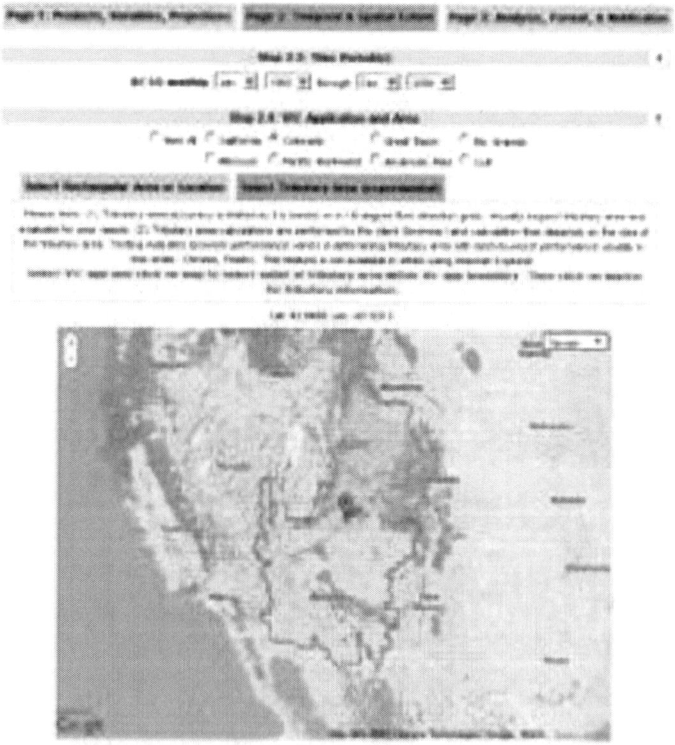

The **Bias Corrected and Downscaled WCRP CMIP3 Climate and Hydrology Projections** computer application provides a gridded archive of 112 contemporary projections of climate and hydrology on a monthly basis. The information covers the western United States for the period 1950 to 2009. The application can be accessed online at: https://gdo-dcp.ucllnl.org/ downscaled_cmip3_projections.

Basin Studies

Reclamation began funding and collaborating with non-Federal entities on Basin Studies in 2009. The elements of each basin study enable the study partners to understand the water supply and demand imbalances in a watershed and also find ways to bring supply and demand into balance. The first element concerns supply and demand – how much water is currently used and expected to be used in the future compared to how much is available to supply those uses. Second, the studies help Reclamation and its partners comprehend the risk of continued imbalances to water and power operations. Third, Reclamation's partners will use this information to improve and optimize operations of existing water supply infrastructure, investigate the need for new infrastructure, recommend institutional reforms, and find ways to reduce demands through conservation and efficiency. Although the Basin Study Program does not fund project construction, the program provides an opportunity for the Basin Study partners to develop coordinated strategies for meeting common water management challenges.

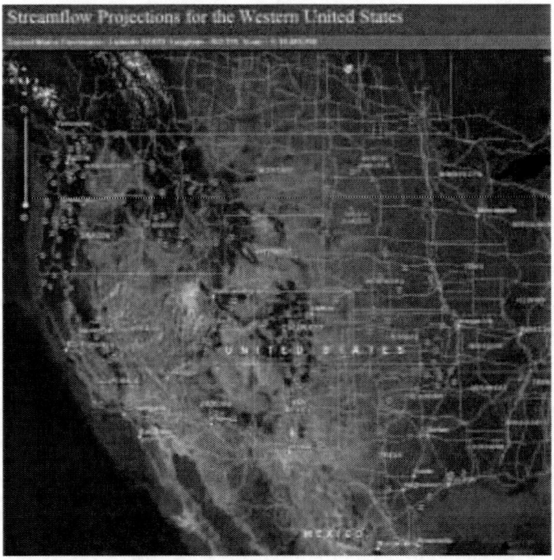

Reclamation's **Streamflow Projections for the Western United States** allows users to access daily and monthly streamflow projection data at 195 locations along streams and rivers in the western United States. The information is based on 112 contemporary climate projections and covers the period 1950 to 2099. The application can be accessed online at: http://gis.usbr.gov/Streamflow_Projections.

Reclamation and its partners are conducting Basin Studies throughout the American West to evaluate water supply and demand imbalances and define options for meeting future water demands.

Each of these elements is examined in the context of a changing climate to prepare for the potential effects of warmer temperatures and other variations in climatic conditions on water supply and demand. To accomplish this task, the Basin Studies build on baseline risk and impact assessments developed through the West-Wide Climate Risk Assessments. Reconnaissance-level information from the West-Wide Climate Risk Assessments provides a consistent basis for more in-depth analyses of river basins in the West. As a result, Basin Studies contribute to a more informed evaluation of possible adaptation options for meeting future water demands and careful consideration of the consequences of various actions on operations planning and other activities.

KEY ELEMENTS OF A WATERSMART BASIN STUDY

- State-of-the-art projections of future supply and demand by river basin.
- An analysis of how the basin's existing water and power operations and infrastructure will perform in the face of changing water realities.
- Development of options to improve operations and infrastructure to supply adequate water in the future.
- Recommendations on how to optimize operations and infrastructure in a basin to supply adequate water in the future.

Reclamation contibuted $12.5 million in funding for 17 basin studies in the first four years of the program. Although the minimum required cost-share from non-Federal partners is 50 percent, the actual non-Federal cost-share exceeds that amount, with non-Federal partners contributing as much as 70 percent of the total cost of a study. Because Reclamation's partners participate at this higher cost-share level, Reclamation has been able to leverage $12.5 million in Federal funding against $13.8 million in non-Federal funding to advance these vital watershed studies. Four of the seventeen studies funded through WaterSMART are described in greater detail below and on the pages that follow to highlight the importance of collaboration to understanding water management issues.

Colorado River Basin Study

The Colorado River basin is vast. Within the United States, the region includes the seven Basin States of Arizona, California, Colorado, Nevada, New Mexico, Utah, and Wyoming. Water from the Colorado River serves a number of beneficial human uses: increasing the reliability of water supplies for municipalities, industries, and farms and ranches; providing water for hydroelectric power generation; reducing the risks of flooding; and providing for recreation. Water from the river also provides an array of environmental benefits, supporting a wide diversity of fish and wildlife and their habitats, maintaining good water quality, and preserving flow and water-dependent ecological systems.

A recent period of prolonged drought in the region prompted the seven Basin States to work with Reclamation to develop strategies for lessening the impacts of water shortage. In 2007, the Secretary of the Interior issued interim guidelines for coordinating the reservoir operations of Lake Powell and Lake Mead – the two largest reservoirs on the river – through 2026 and to more effectively manage water supplies during periods of shortage within the lower Basin States. Interior worked closely with the States, Native American Tribes and communities, non-governmental organizations, and many other stakeholder groups to develop the guidelines.

The collaborative process strengthened relationships among the stakeholders, producing a regional partnership capable of responding to persistent drought.

With the interim guidelines in place, Reclamation and its partners engaged in additional long range planning to further study the problems caused by water supply and demand imbalances in the basin. One of those studies, the Colorado River Basin Study, began in January 2010 as part of the WaterSMART Basin Study Program. The purpose of the study is to define the extent of water supply and demand imbalances in the basin through the year 2060. The study will also develop and analyze strategies to resolve those imbalances under a range of conditions that could occur during the next 50 years. The study is being conducted by Reclamation and seven States in collaboration with a broad range of stakeholders including Native American tribes and communities, agricultural users, purveyors of municipal and industrial water, power users and providers, recreational groups, conservation groups, and the general public.

Visible rings along the shoreline of Lake Powell due to prolonged drought.

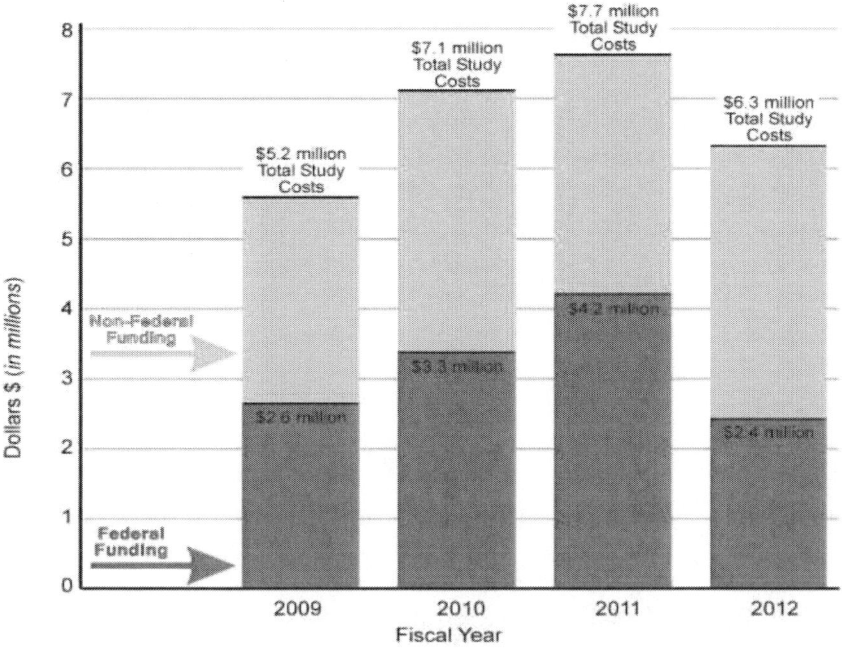

Basin Study Funding Since 2009. The Bureau of Reclamation has leveraged $12.5 million in Federal funding to conduct over $26 million in collaborative water resources planning through the Basin Study Program.

> "Through the Colorado River Basin Study, the Basin States have taken established partnerships to a higher level, providing huge benefits to the States."
>
> —Kay Brothers, Deputy General Manager of Engineering and Operations (retired) Southern Nevada Water Authority

Of particular importance, the Colorado River Basin Study provided the study participants an opportunity to collaborate on scenario planning. Given the historical variability of Colorado River inflows and the potential for increased variability in the future, there is great uncertainty associated with future water supply throughout the basin over the next 50 years. Developed with input from a broad range of Basin Study participants, the water supply and demand scenarios incorporate key factors that will drive future uncertainty. These factors include changes in natural systems including climate change, changes in demographics and land use, changes in technology and economics, and changes in social and governance structures.

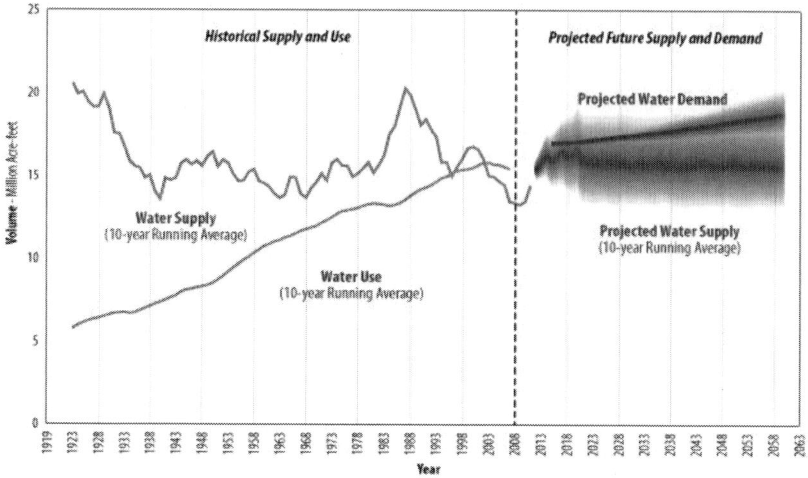

Notes:

Water Supply represents natural flow as measured at the Colorado River above Imperial Dam, Arizona

Water Use and Demand include deliveries to Mexico in accordance with the 1944 Treaty with Mexico and losses such as those due to reservoir evaporation, and operational inefficiencies.

Projected Water Supply is computed as the average 10th, 50th (median), and 90th percentiles of the Study's 4 water supply scenarios. The average of the medians is indicated by the darker shading.

Projected Water Demand is represented by the Study's 6 water demand scenarios. The median of the scenarios is indicated by the darker shading.

Historical supply and use and projected future Colorado River Basin water supply and demand.

Needs for water in the basin and the availability of that water differs as dramatically as the region's landscape. Consequently, the magnitude and timing of projected water supply and demand imbalances vary significantly across the basin. Preliminary results show the imbalance may be as high as 3.5 million acre-feet over the next 50 years, particularly when considering potential changes in climate. This large variance in projected imbalances will require the study partners to consider a wide variety of options to pursue a secure water future for the region.

Although the study will not result in the selection or funding of a particular proposed option, it is a critical first step toward establishing the common technical foundation for coordinated strategies to reduce future water supply and demand imbalances. Kay Brothers of the **Southern Nevada Water**

Authority and Ted Kowalski of the **Colorado Water Conservation Board** stress the importance of the partnership to achieving workable solutions. Through collaboration on Colorado River Basin Study, the study partners are working cooperatively, strengthening the work and capabilities of all the partners.

Input on how to help resolve future water supply and demand imbalances has been sought from stakeholders with an interest in the basin. The study partners received over 150 ideas that were organized under the categories of increased supply, reduced demand, modification of operations, and governance and implementation.

> "The ability to share the collective wisdom and experience of water managers representing the seven Basin States,Tribes, the Federal government, and NGOs benefits all of the states involved."
>
> —Ted Kowalski, Interstate and Federal Section Chief Colorado Water Conservation Board

The first study report was released in June 2011 and additional technical updates have been released since that time to reflect study progress and the ongoing input of stakeholders. A final report due out in the fall of 2012 will summarize the study, including findings related to the performance of various representative options and strategies. The study will create a framework for future collaborative discussions and planning efforts related to water management in the basin where potential changes in climate, record drought, population increases, and environmental needs have heightened competition for scarce water supplies.

Through the Colorado River Basin Study, Reclamation, the seven Basin States, and other interested stakeholders have worked collaboratively to examine supply and demand imbalances on the system from a basin-wide perspective. The preliminary results clearly show that such imbalances exist now and will persist into the future. Managing shared resources presents challenges that must be answered cooperatively. The Basin Study Program provides a forum for seeking cooperative solutions and the broad stakeholder involvement featured in the Colorado River Basin Study is a unique example of the benefits of the program.

Yakima River Basin Study

The Yakima River Basin Study, the first one completed under the WaterSMART program, continued an existing partnership between

Reclamation and the Washington State Department of Ecology. Reclamation and the State are now using the information from the basin study to develop a resources management plan that integrates water needs.

Through an innovative and progressive take on partnership and collaboration, the Yakima River Basin Water Enhancement Project (YRBWEP) group sought to create a plan of action to solve decades of chronic water shortages worsened by drought. The water of the Yakima River basin is vital to a wide and diverse array of water users and needs. The YRBWEP group represents that diversity of needs by including irrigators, growers, municipalities, State and Federal agencies, Native American Tribes, and non-government organizations. This group met every two weeks for several months beginning in 2009 to develop a mutually acceptable plan for managing water in the Yakima River basin. The plan includes several critical elements the State of Washington will use to increase the reliability of water deliveries while improving the river ecosystem, including enhanced water conservation, new or expanded water storage, groundwater storage, water marketing, modifications to existing operations and facilities, fish passage, and fish habitat restoration.

Reclamation and its study partners contributed to the collaborative effort by combining previously-developed studies of water operations in the basin with climate projection data to create a Basin Study model. The model provided the information necessary for the group to consider the impacts of changing climate on those seven critical elements and to understand in greater depth the range of future water supply and demand imbalances.

The American Water Resources Association recognized the efforts of this broad partnership of Federal, State, Tribal, and non-governmental stakeholders with the association's first Integrated Water Resources Management award in 2012.

Bumping Lake in Washington State.

The Yakima River Basin Water Enhancement Project: Partnership and Progress

The YRBWEP Action Plan

- Enhanced water conservation
- New or expanded water storage
- Groundwater storage
- Water marketing
- Modifications to existing operations and facilities
- Fish passage
- Fish habitat restoration

Milk and St. Mary Rivers Basin Study

Also recently completed, the Milk and St. Mary Rivers Basin Study is another example of the benefits of scenario planning. All previous models for the Milk and St. Mary Rivers system relied on an assumption that future hydrology would look much like the hydrology of the past – a condition termed "stationarity." The models developed through the Basin Study Program included this traditional historic conditions model as well as five additional scenarios to assess the effects of various warming trends. The models developed for this basin study operate at a finer time-step resolution than previous models did. The results from all scenarios modeled showed increased shortages. Surprisingly, the shortage was not due to a decrease in water supply. Instead, the trend of future warming resulted in longer growing seasons and greater crop demands on water. While the Milk River Project is expected to continue delivering a volume of water consistent with current amounts, the increased demands will worsen shortages in the basin.

Aerial view of the Fresno Reservoir on the Milk River in North Central Montana.

While the Milk River Project was authorized for irrigation purposes, it also provides a number of ancillary benefits, such as water that enhances fish and wildlife habitat. As demand on authorized purposes increases and as efficiencies in irrigation delivery and application increase, the secondary benefits will be harder to maintain.

The Federal government operates the major irrigation delivery infrastructure in the region, so Reclamation's involvement is essential to State water management. Federal involvement is also needed because of issues related to Native American trust responsibilities and international waters. Reclamation has a primary role in working with the Tribes and the State to implement water rights compacts in ways that achieve the goals of the compacts for the Tribes, while also protecting other water uses.

Reclamation provided the expertise needed to develop climate change scenarios. The State's in-house hydrologist collaborated with Reclamation's climate scientists and modelers to run the future climate scenarios with the river system model. The model will improve negotiation and implementation of Tribal compacts. The model will improve the State's ability to evaluate international water sharing arrangements with Alberta, Canada. The State will use the model as part of a decision support system to make informed decisions on how to direct resources to the best options.

Water flowing to Lake Bowdoin National Wildlife Refuge in North Central Montana.

"Basin modeling enabled us to understand how water supplies in the future, although they might be similar to what we have today, will not be sufficient to meet future needs in the region for irrigation and other uses. Warming trends will mean higher shortages in the future due to a longer growing season and higher crop irrigation demands for the water that we have available to us."

—Larry Dolan, Hydrologist Montana Department of Natural Resources

The Basin Study confirmed the State's imperative to make faster progress on infrastructure improvements. Reclamation and the State will continue to use the model to assess water management related to endangered species and habitat issues, such as for the threatened bull trout.

Santa Ana River Watershed Basin Study

The Santa Ana River Watershed Basin Study illustrates how effective collaboration can improve water management. This basin study takes a crest-to-coast and corner-to-corner approach throughout a 2,400 square mile watershed, which encompasses everything from beach to mountain communities, to address the area's unique water resource challenges. The Santa Ana River watershed is home to more than 6 million people, and tens of millions more people visit the area each year. The region also has a large manufacturing, industrial, and agricultural base. Over 350 water, wastewater, and groundwater management, flood control, environmental and other non-governmental organizations, are working together as part of the Santa Ana River Watershed Project Authority on an Integrated Regional Water Resources Management Plan process called the "One Water One Watershed" Plan, or OWOW. Through the Basin Study Program, Reclamation is able to work in partnership with the Authority and its stakeholders to update key components of the plan.

"The basin study provides a forum for a shared experience that engenders trust."

—Celeste Cantú, General Manager Santa Ana Watershed Project Authority

The OWOW Plan takes a region-wide look at water supply and demand, population growth, water quality, flood control, and projected effects from climate change. Local water sources like the Santa Ana River are critical to the region, but water imports from the Colorado River and State Water Project also play a key role. However, by the time imported water reaches arid southern California, water from the Colorado River contains high concentrations of salt and many of the region's agricultural and manufacturing industries generate salt that requires disposal. This salt imbalance is just one example of the many challenges facing local water managers who work collaboratively through the Authority to effectively and efficiently manage water in the region.

Celeste Cantú, General Manager of the Authority, recently noted: "Western States and water management agencies need Reclamation's skills because Reclamation is a Federal agency whose staff understand best the water, culture, and needs here in the West." Reclamation and the Authority share a vision for effectively managing the Santa Ana River basin's finite water resources to meet future needs and are working in partnership to assess past, present, and future water needs. The Santa Ana River Watershed Basin Study is helping the Authority identify data gaps, conduct trade-off analyses, address the effects of climate change, and develop effective adaptation strategies.

Celeste Cantú, General Manager of the Santa Ana Watershed Project Authority, and Jack Simes, Area Planning Officer for Reclamation's Southern California Area Office, at the Santa Ana Watershed Basin Study Climate Change Workshop.

Landscape Conservation Cooperatives and Climate Science Centers

Interior established the Landscape Conservation Cooperatives and Climate Science Centers as collaborative institutions to promote effective resource management based on representation of multiple scientific disciplines, an emphasis on consensus and shared decision-making authority, and reciprocal cooperation that increases trust, reduces conflict, and achieves common goals.

Climate Science Centers provide scientific information, tools, and techniques that land, water, wildlife, and cultural resource managers and other interested parties can apply to anticipate, monitor, and adapt to climate and ecologically-driven responses at regional and local scales. Climate Science Centers prioritize the delivery of fundamental science, data, and decision-support activities to meet the needs of the on-the- ground managers. This includes working with members of the Landscape Conservation Cooperatives and others to provide climate change impact information on natural and cultural resources and to develop adaptive management and other decision-support tools for managers.

Landscape Conservation Cooperatives are independent partnerships that involve Federal, State, Tribal, non-governmental conservation organizations, and other partners. They have smaller geographic areas than the Climate Science Centers. Thus, they play an important role in identifying the priorities for science needed most by local resource managers. Because the areas for Landscape Conservation Cooperatives are smaller than those for Climate Science Centers, each Climate Science Center solicits science needs from multiple Landscape Conservation Cooperatives and other regional management stakeholders. These will be reconciled to create a set of regional priorities by the Climate Science Center's stakeholder committee, which will include representatives from relevant Landscape Conservation Cooperatives and other management entities. This strategic set of science needs will be captured in a science plan at each Climate Science Center, which will form the basis for annual project selection.

Landscape Conservation Cooperatives and Climate Science Centers will also work closely together on the conduct and delivery of science. It is now clear that effectively providing scientific information for management involves ongoing and direct engagement between scientists and resource managers.

Access Reclamation's SECURE Water Act Report to Congress at: www.usbr.gov/climate

SECURE Water Act Report to Congress

In 2011, the Secretary of the Interior issued the first *SECURE Water Act Report to Congress*. Using information generated from Reclamation's West-Wide Climate Risk Assessments, the report depicts the potential effects of climate change in the western United States. The report provides quantitative assessments of climate change impacts on temperature, precipitation, and streamflow. It also qualitatively describes the range of effects that climate change is projected to have on water supply from surface water, groundwater, and snowmelt, on reservoir operations, on flood management, and on the fish and wildlife in eight major basins of the western United States.

According to the report, projections of climate change and resulting impacts to water supply and demand vary across different parts of the western United States. Warmer temperatures may result in less snowpack accumulation in mountain areas, which will in turn change the amount and timing of snowmelt and other precipitation runoff. Changing climate conditions may also result in shifts in the amount and duration of rain received during the summer months. As a result, changes in climatic conditions are projected to have a wide range of effects on supply and demand due to shifting rain and snow patterns, winds, and temperatures.

The risks of flooding in the summer months appear to be greater under climate change. And in the absence of increased precipitation, agricultural water demands may increase significantly because plants will use more water during longer growing seasons. Subsequent editions of the report will continue to provide the information Congress and the American public need to better understand these kinds of changes as well as the range of other effects that climate change is projected to have on water supply operations, on the generation of electricity at hydropower facilities, and on fish and wildlife and their habitats.

Next Steps

New Frameworks for the Basin Study Program

SECURE feasibility studies, the final phase of the Basin Study Program, will provide Reclamation and its partners the opportunity to find ways to take action to bring water supply and demand into balance. After four years, significant progress has been made in the Basin Study Program. Reclamation published a framework with program guidelines in 2009. Incorporating lessons learned from four years of experience with studies such as the ones described in this report, Reclamation will finalize its internal directives in 2013. Moreover, with three studies completed and several more in the final stages of completion, Reclamation developed a framework to help the non-Federal partners take the next step.

Feasibility studies are more detailed investigations of the technical, economic, and environmental viability of a wide range of possible actions. Basin studies help Reclamation and its partners project the demand for water in the future, the availability of water to meet that demand, and the effects that climate change has on both. Options will include mitigation actions – those actions that reduce stress on water supply systems – developed through the basin studies. Because some change is inevitable, options will also include adaptation actions to sustain the resiliency of ecosystems and the dependability of water supply systems.

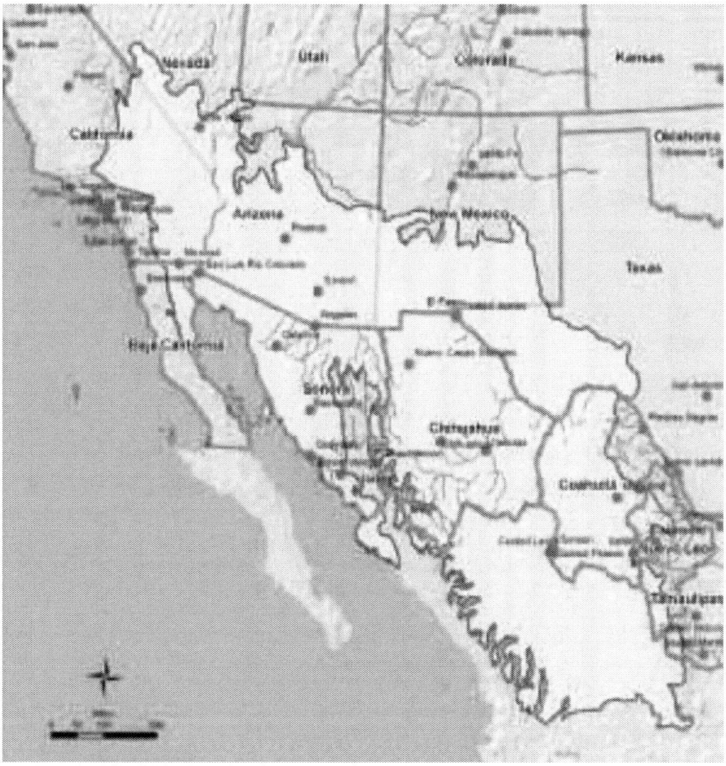

Geographic boundaries of the Desert Landscape Conservation Cooperative.

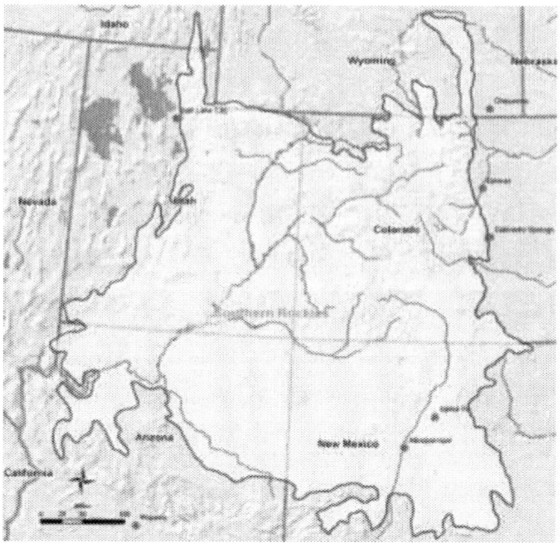

Geographic boundaries of the Southern Rockies Landscape Conservation Cooperative.

SECURE feasibility studies will not include recommendations regarding construction authority or appropriations. Instead, recognizing that governments at all levels face budget constraints and competing priorities for limited funding, Reclamation's role in undertaking these studies is to work with its partners to collaboratively develop regional or basin-scale solutions. As States, Tribes, and local entities continue to take greater leadership roles in water resources investments, Reclamation expects that project sponsors will see value in partnering with the bureau to complete feasibility studies.

Reclamation expects to publish a framework establishing the requirements of a pilot SECURE feasibility study in 2012 and announce the award of a pilot study in 2013. The pilot study will help Reclamation evaluate ways to incorporate climate change into the planning process. Planning for climate change through SECURE feasibility studies contributes to targets under Interior's Priority Goal for Climate Change, which provieds an assessment of the Nation's vulnerability to climate change and identifies coordinated adaptation response actions.

Activities of the Desert and Southern Rockies Landscape Conservation Cooperatives

The Desert Landscape Conservation Cooperative recently published a report outlining the group's accomplishments to date, as well as ongoing and planned work. Through significant outreach efforts combined with a review of scientific and technical documents, the Science Working Group identified 533 research topics to improve our knowledge of the interdependence of ecosystems and hydrology. This research will tell us how ecosystems and hydrology interact to support life in the desert and how climate change will impact that life support. The Steering Committee established a set of criteria to focus research efforts on the most critical gaps in scientific and technical understanding of the desert ecosystems in the Western U.S. Because Landscape Conservation Cooperatives support the application of basic science to resource management, the criteria emphasize and prioritize the science projects that will be most useful to managers in the field. Through this effort, the group has been able to disseminate information about progress on high priority research questions and spotlight needs for additional new research in areas of applied science.

The Southern Rockies Landscape Conservation Cooperative Steering Committee also developed priority resource areas for research. The group identified several broad categories of needed applied science research into the forests, river systems, and plant and animal species either endangered or critical to the healthy and sustainable functioning of ecosystems in this unique and diverse region of the country. Using prioritization criteria tailored to the environmental resource issues of the Southern Rocky Mountain region, the Science Working Group developed a list of six high priority science needs and five coordination needs.

Desert Landscape Conservation Cooperative Steering Committee

Arizona Game and Fish Department	Natural Resources Conservation Service
Bureau of Indian Affairs	Nevada Department of Wildlife
Bureau of Land Management	New Mexico Office of the State Engineer
Bureau of Reclamation	Rio Grande Joint Venture
California Association of Four Wheel Drive Club	Rocky Mountain Bird Observatory
California Department of Fish and Game	Sky Island Alliance
Cocopah Indian Tribe	Sonoran Joint Venture
Colorado River Indian Tribes	Texas Parks and Wildlife Department
Fort McDowell Nation	U.S. Department of Defense
Gila River Indian Tribe	U.S. Fish and Wildlife Service
Instituto Nacional de Ecología	U.S. Forest Service
National Park Service	U.S. Geological Survey
Native American Land Conservancy	

In 2013, the Southern Rockies Landscape Conservation Cooperative will participate in an integrated and organized effort together with the Great Northern, Plains and Prairie Potholes, and Great Basin Landscape Conservation Cooperatives to support planning efforts for greater sage-grouse.

In 2013, the Southern Rockies Landscape Conservation Cooperative will participate in an integrated and organized effort together with the Great Northern, Plains and Prairie Potholes, and Great Basin Landscape Conservation Cooperatives to support planning efforts for greater sage grouse. The focus of the collaboration is to engage all four Landscape Conservation Cooperatives overlapping sage grouse habitat to locate, consolidate, analyze and provide a unified point of access for sage-grouse data to all those engaged in planning a successful sage-grouse conservation strategy. The specific outcomes of this three- year effort will include: 1) a clearinghouse for sage-grouse population and sagebrush ecosystem data; 2) an applied analysis of future habitat condition and population vectors useful to guide conservation planning; and 3) decision support tools to assist management agency partners to make science-informed decisions about allocating resources for grouse conservation. Funding for this effort is being provided by the Fish and Wildlife Service, using Adaptive Science funding provided by the Fish and Wildlife Service Office of Science Advisor.

The Southern Rockies Landscape Conservation Cooperative is also working with the Bureau of Land Management (BLM) to coordinate efforts related to BLM's Ecoregional Assessments and the Southern Rockies Landscape Conservation Cooperative's comprehensive science assessment. The BLM has requested the Southern Rockies Landscape Conservation Cooperative to host workshops to allow the Landscape Conservation

Cooperative Science Working Group to review the Colorado Plateau Rapid Ecoregional Assessment report and facilitate drafting of an interagency regional report summarizing its findings, opportunities and challenges. The Southern Rockies Landscape Conservation Cooperative intends to use these same workshops to support the development of a comprehensive science needs analysis for the Landscape Conservation Cooperative, also planned for FY 2013. Through this collaborative effort, the Southern Rockies Landscape Conservation Cooperative and the BLM can share information about their respective efforts, avoiding duplication and avoiding partner fatigue by addressing two purposes through the workshops.

Reclamation recently announced the award of $1.65 million in cost-shared WaterSMART funds for 19 new Landscape Conservation Cooperative applied science projects in 2012. The projects were selected through a competitive process. Through these partnerships with States, Tribes, water users, research institutions, and nonprofit groups, Reclamation is promoting shared learning while advancing progress on the priority science needs of the Desert and Southern Rockies Landscape Conservation Cooperatives. Working in concert with Reclamation, the Fish and Wildlife Service also funded several projects in high-priority research areas. To maximize the utility of its WaterSMART research funding, the Fish and Wildlife Service focused its financial resources on complementary investigations into areas not already funded through Reclamation's program. Funded research areas included aquatic habitat restoration for desert ecosystems and an evaluation of climate change impacts on birds of the Sonoran Desert.

Southern Rockies Landscape Conservation Cooperative Steering Committee

Arizona Game and Fish Department	New Mexico Office of State Engineer
Bureau of Indian Affairs	Pueblo of Jemez
Bureau of Land Management	Shivwits Band of Paiutes
Bureau of Reclamation	The Nature Conservancy - Migratory Bird Program
Colorado Department of Natural Resources	Trust for Public Lands
Colorado Natural Heritage Program	U.S. Fish and Wildlife Service
Intermountain West Joint Venture	U.S. Forest Service
National Park Service	U.S. Geological Survey
Natural Resources Conservation Service	Utah Governor's Public Lands Policy Coordination Office
New Mexico Department of Game and Fish	Wyoming Game and Fish Department

Progress on the nine projects awarded in 2011 continues. The preliminary findings of those studies promise significant improvements in our understanding of ground and surface water interactions, fish habitats and species survival mechanisms, and the biodiversity of fish in waters common to the U.S. and Mexico.

THE ENERGY-WATER NEXUS

Water and energy are inextricably linked. Water is necessary to generate power, whether it is the kinetic source to turn a hydroelectric turbine; the thermal workhorse for a coal-fired power plant, cooling, and emissions control; or the water needed for mining, transportation, and decommissioning for other electricity-generating sources. Likewise, energy is a necessary component of water. Energy is required to store, move, and treat water on the supply side as well as the waste side. According to the California Energy Commission's 2005 Integrated Energy Policy Report, the State of California's annual water-related energy accounts for about 20 percent of the State's total electricity consumption, about a third of non-power plant natural gas consumption, and about 88 million gallons of diesel fuel consumption. The Oregon Water Resources Department estimates that about 21,000 pumps were used in 2008 to pump irrigation water in Oregon, resulting in electricity costs of about $49 million that year. In this sense, water saved is energy earned, and energy saved is water earned.

While Reclamation's earliest vision was to help reclaim the arid west, primarily through irrigation of land to make it agriculturally productive, Interior and Reclamation have long recognized the value of developing hydroelectric power as we strive to extract the most out of each drop of water. Today, Reclamation is the Nation's second largest generator of hydroelectric power, operating 53 hydroelectric power plants with a capacity of about 15,000 megawatts of electricity per year. Reclamation generates an average of 40 billion kilowatt-hours each year—enough to power almost 3.5 million homes.

Power from Reclamation facilities is transmitted on the interconnected transmission grid and sold on the wholesale power market. Bonneville Power Administration and Western Area Power Administration, which are part of the Department of Energy, market Reclamation's power to customers. Reclamation, along with Bonneville and Western, serve customers in 17 western and central States.

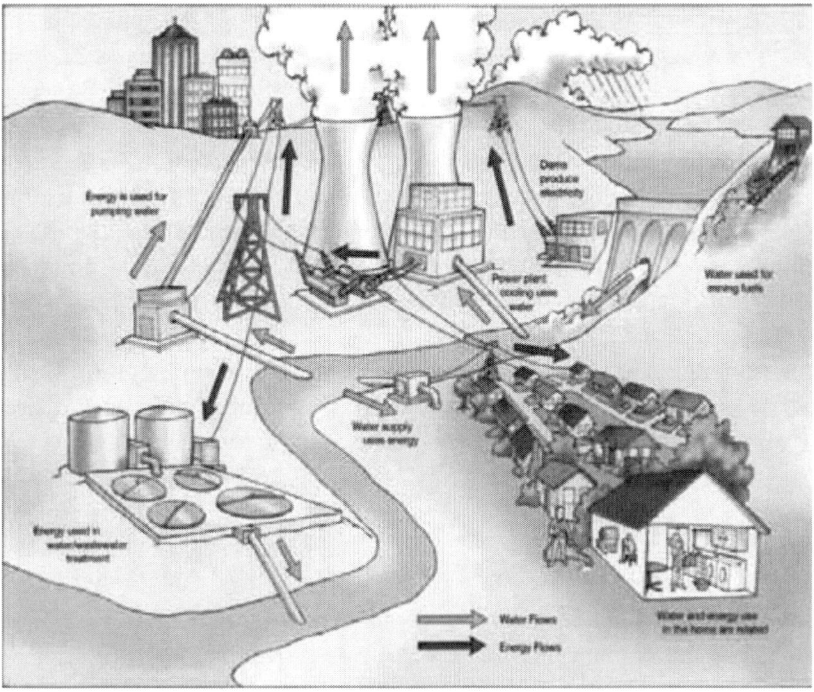

The relationship between energy and water, from the United States Department of Energy's 2006 report, **Energy Demands on Water Resources: Report to Congress on the Interdependency of Energy and Water**.

The Ponderosa Powerplant, installed by Swalley Irrigation District in Oregon with ARRA funding (top right). The powerplant, part of a project to increase the use of renewable energy and also address water efficiency, is housed in a building adjacent to an existing irrigation canal to take advantage of a drop in elevation at the site (above).

In 2010, as Reclamation expanded existing programs and incorporated them into WaterSMART, several steps were taken to emphasize the link between energy and water—each in consideration of the other. Projects that implement renewable energy are now eligible for WaterSMART Grant funding. In its evaluation of funding proposals, Reclamation also prioritizes those proposals that describe the estimated energy savings from planned water efficiency improvements. These projects have included everything from installing new hydroelectric turbine generators on canals and conduits; to building storm water recharge systems—taking advantage of local water, thus minimizing the need to pump water from distant sources; to building wind power generation— replacing supplied power for pumping; to installing automated systems on facilities—avoiding driving and fuel consumption; to increasing availability of water for downstream hydroelectric facilities.

Progress to Date

Since 2010, Reclamation has awarded six WaterSMART Grants for projects that incorporate renewable energy improvements into existing water delivery facilities. Together, those projects are expected to result in over 3500 kilowatts of new renewable energy capacity.

Through an FY 2011 WaterSMART Grant, the **Boise Project Board of Control**, near Homedale, Idaho, west of Boise, is developing an 839 kilowatt power plant at the "Fargo Drop," a change in elevation in the existing irrigation canal system. The project will generate hydroelectric power that will be sold to the Idaho Power Company, and revenues are expected to offset as much as 30 to 40 percent of the operational costs of the irrigation system. The project will also include installation of a Supervisory Control and Data Acquisition system to improve regulation flows in the Deer Flat Low Line Canal below the Fargo Drop diversion. The new power plant is expected to be operational in the spring of 2013.

This year, another project in Idaho that incorporates renewable energy improvements into the delivery system has been selected for WaterSMART Funding. The **Consolidated Irrigation Company** in the Bear River basin in southern Idaho will convert an unlined earthen canal to high-pressure pipe and install advanced measuring devices, improvements expected to result in water savings of over 9,400 acre-feet each year once the project has been completed. To take advantage of an elevation change as the new pipeline drops into Glendale Reservoir, the project also includes the installation of a 500-kilowatt hydropower facility to generate renewable energy. The project is expected to begin construction in 2013.

Sponsors of WaterSMART Grant projects continue to explain how their proposed water efficiency improvements can be expected to lead to energy savings as well. Since 2010, Reclamation has provided funding for more than 60 water efficiency projects whose sponsors provided energy savings estimates. Those projects together are expected by their sponsors to save over 40 million kilowatt-hours annually, the amount of energy necessary for about 3,400 households. Sponsors have used a number of different methods and assumptions to estimate energy savings:

The **City of Torrance, California**, has been selected to receive a WaterSMART Grant in 2012. The City will make improvements so that storm water can be filtered and used to recharge groundwater rather than flowing untreated to the ocean. The City estimates that by using this local source of water over the twenty-year life of the improvement, about 1,500,000 kilowatt-hours of energy necessary to deliver an equivalent amount of water from outside the region could be avoided. To make that estimate, the City drew on existing estimates that 3,000 kilowatt-hours of energy are required to move each acre-foot of water through the State Water Project to southern California.

> "The WaterSMART Grant has made the difference in making this a more viable project for us."
>
> —Ken Grover Boise Project Board of Control

The **Southern Nevada Water Authority**, a 2011 WaterSMART Grant recipient, took a slightly different approach to formulate an energy savings estimate. The project is expected to generate 790 acre-feet of water savings per year through rebates to residential property owners for replacement of turf grass with water efficient landscaping. To estimate energy savings that could result from the project, the Authority considered the energy necessary to treat and deliver water to the average customer: each acre-foot of water saved is expected to yield an estimated 2,118 kilowatt-hours in energy savings. The Authority expects to save about 1,600,000 kilowatt-hours annually using that approach.

Some applicants have quantified energy savings that will result from a reduction in pumping associated with water savings. The **South Board of Control in Owyhee, Idaho**, a 2011 WaterSMART Grant recipient, is converting open lateral canals to closed-pipe systems that will supply gravity-flow irrigation water to farm units, conserving an estimated 3,300 acre-feet of water annually. Currently, nine pumps supply water to three canals to meet supplemental needs, consuming 18 to 20 million kilowatt-hours of energy annually. The Board of Control worked with the Bonneville Power Administration to calculate that avoided pumping of 3,300 acre-feet of water each year will result in savings of 573,911 kilowatt- hours of energy annually.

Other applicants have quantified reductions in vehicle miles driven and reductions in carbon emissions expected to result from water efficiency improvements. Through a WaterSMART Grant, the **Shasta Community Services District** near Redding, California, is implementing a project to allow remote monitoring and control of its water distribution system. The District estimates that it will avoid about 2,080 vehicle miles annually, or about 149 gallons of fuel, by reducing the number of on-site visits by staff. Additionally, some project sponsors have incorporated more energy-efficient equipment into their water delivery systems. The **Whitestone Reclamation District** in Washington, a 2011 WaterSMART Grant recipient, is installing a new high-efficiency booster pump at its pumping plant. Based on the average number of hours that the pump runs and the current pumping plant efficiency, installation of the high- efficiency pump is expected to save 12,343 kilowatt-hours of energy per year.

In 2010, Reclamation also restructured its WaterSMART System Optimization Review funding criteria to prioritize proposals that assess ways to increase the use of renewable energy and improve energy efficiency. As part of a System Optimization Review, the **Southwest Kansas Groundwater Management District No. 3**, in Garden City, Kansas, is examining ways to increase both water and energy efficiency in its use of Arkansas River water supplies, including the potential for installing low-head hydroelectric facilities, wind powered headgates and solar powered headgates.

Similarly, as part of new funding criteria incorporated into the Title XVI program in 2010, Reclamation prioritizes proposals based on the extent to which each project incorporates renewable energy or addresses the energy required for treatment and delivery of recycled water.

Secretarial Order 3297, which guides the WaterSMART Program, is broad in its charge to integrate water and energy policies to support sustainable use of all natural resources. In addition to the water and energy components of WaterSMART specified above, Reclamation leverages its sustainable water and power influence through other programs. In March 2010, the Departments of the Interior, Energy, and Army entered into a memorandum of understanding to help meet the Nation's hydropower needs and to align ongoing and future renewable energy efforts.

In April 2012 these agencies reported on progress, including the following accomplishments:

- Reclamation has identified 373 existing canals and conduits and 191 existing dam and reservoir sites with the total combined potential of generating an additional 1.565 million megawatt-hours annually. Additional details are available at: www.usbr.gov/power/CanalReport/ and www.usbr.gov/power/AssessmentReport/USBRHydroAssessmentFinalReportMarch2011.pdf
- To assist hydroelectric developers in responding to these resources, Reclamation is drafting Reclamation Manual Directives and Standards for Lease of Power Privilege Processes, Responsibilities, Timelines, and Charges, which will guide applicants through a streamlined process for installation of power generation on existing facilities.
- Reclamation is actively involved in the Basin Scale Opportunity Assessment to develop an approach to hydropower and environmental assessment, emphasizing sustainable energy systems and environmental protection and restoration. http://basin.pnnl.gov
- Overall, since 2009, Reclamation has worked with its partners to install 77.5 megawatts of new hydropower generating capacity through power plant upgrades. In addition, 35.9 megawatts of capacity have been developed through the Lease of Power Privilege and Federal Energy Regulatory Commission licensing processes, with significant amounts of additional capacity under active consideration.

Reclamation is installing hydropower optimization systems for plant operators to identify and gain efficiencies. At most Reclamation facilities, a fixed water release dictates the available power capacity at any given point in time. By optimizing the plant to maximize the power capacity based on plant conditions such as efficiency curves, rough zones, outages, ancillary service demand, and the scheduled water release, the energy per acre foot of water will also be maximized. This is tantamount to increasing plant efficiency through conventional methods, such as generator rewinds and turbine uprates. It can also offset the need to build new renewable power facilities such as new hydropower plants, wind farms, or solar arrays, all of which have significant environmental and budget impacts. Optimization adds extra generation capacity to the power system without expending any fuel or water. Reclamation estimates that this system can give operators the real-time information to improve operations by one percent across Reclamation, producing 16.2 megawatts of additional power at all times.

Through WaterSMART, Interior is taking active steps to assist its partners as they consider the relationship between energy and water in planning and implementing their projects and operations. Reclamation is taking advantage of the energy capacity available on existing facilities and updating facilities to be more efficient. These and other steps are key as we work together toward sustainable management of water and energy resources.

National Water Availability and Use Assessment

A key part of achieving Interior's sustainability goals is to inform the public and decision makers about the status and changes over time of the Nation's freshwater resources. To achieve these goals, the WaterSMART program has developed plans and begun implementation of a National Water Census by the USGS. The Census will provide a more accurate picture of the quantity and quality of the Nation's water resources for beneficial uses and the basis for an improved ability to forecast the availability of water for future economic, energy production, and environmental uses. The concept of a Census is consistent with the SECURE Water Act, which calls for the establishment of a "national water availability and use assessment program" within the USGS.

A national assessment of water availability and use is a complicated undertaking. At the time of the previous national assessments more than 30 years ago, water availability was viewed in comparatively simple terms. These early assessments focused largely on basic statistics about the quantities of water available for various human uses. Since then, competition for water resources has increased considerably and greater importance is attached to water for environmental and ecosystem needs. Likewise, concerns have grown about groundwater depletion, streamflow alteration, climate change and variability, and water-quality impairment. There is also much greater awareness of the connectivity of surface water and groundwater and the linkages between water availability and use of other natural resources.

> "National water availability and use has not been comprehensively assessed in more than 30 years."
>
> —U.S. General Accounting Office (2003)

The Census synthesizes information at the regional and national scales, with an emphasis on compiling this information in a way that is useful to States and others responsible for water management and natural resource issues. Considerable emphasis is placed on Web-based methods for delivery of information. The USGS continues to work with other organizations, Federal and non-Federal agencies, and universities to ensure that the information produced can be aggregated with other types of water-availability and socioeconomic information. To maximize the utility of the information, the USGS coordinates the design and development of the effort through the Federal Advisory Committee on Water Information.

A major undertaking of the Census is to provide estimates of selected water-budget components at consistent spatial and temporal scales across the country. Water budgets account for the inputs, outputs, and changes in the amount of water in components of the water cycle. They are the hydrologic equivalent of the deposits, withdrawals, and change in balance in a checking account and comprise the hydrologic foundation for analysis of water availability. Basic components of water budgets are precipitation, evapotranspiration, surface-water and groundwater flow into and out of the watershed, change in surface-water and groundwater storage, change in snow and ice storage, and human withdrawals and interbasin transfers.

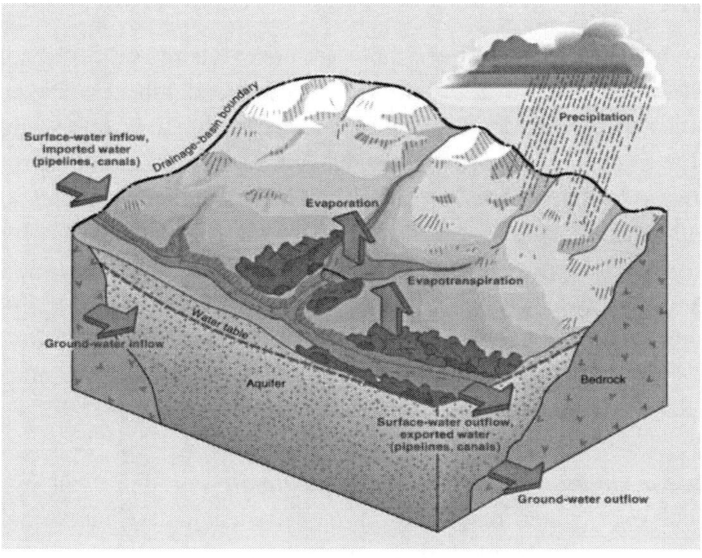

Components of a simple water budget for part of a watershed.

Progress toward Goals

The SECURE Water Act authorized $20 million for each of fiscal years 2009 through 2023 for the National Water Availability and Use Assessment Program. The first appropriation for this effort was $4 million in FY 2011, followed by an appropriation of $6 million in FY 2012. With this funding, the USGS engaged stakeholders in a discussion of priorities. In 2011, the USGS began Geographic Focus Area Studies of water availability and use in the Colorado, Delaware, and Apalachicola-Chattahoochee-Flint River Basins. In addition, five topical areas were identified to enhance future capability for a full Census.

These were (1) estimates of streamflow at ungaged locations and characterization of long-term trends in streamflow; (2) regional groundwater availability studies of principal aquifers; (3) use of remote sensing to quantify evapotranspiration; (4) improved information on human water withdrawals and return flows; and (5) developing tools and web-available resources to understand the effects of streamflow alteration on aquatic ecosystems. We report on progress to date on each of these activities below.

Geographic Focus Area Studies

Throughout the United States there are areas where competition for water resources has reached a level of national attention and concern. The competing interests may arise from multiple human needs (demands for potable water, irrigation, energy, industrial processes, etc.), from competition between human and aquatic ecosystem needs, or both. The Census includes a series of studies, focused on selected large watersheds, where there is a desire on the part of watershed stakeholders to conduct a comprehensive technical assessment of water availability with the best available tools. These Geographic Focus Area Studies contribute toward ongoing assessments of water availability in these watersheds and also provide opportunities to test and improve approaches to water availability assessment.

Apalachicola-Chattahoochee-Flint River Basin

Conflict over water resources in the Apalachicola-Chattahoochee-Flint River Basin among Alabama, Florida, and Georgia has resulted from increases in water use for municipal and industrial supplies, power generation, and agriculture. Competition over water is not limited to the State borders; during drought conditions competition among all water users can become pronounced. The Apalachicola-Chattahoochee-Flint Geographic Focus Area Study will provide information for water managers including enhanced water-use information and linked surface-water and groundwater models. This information also provides input to ecological models that predict changes in fish and mussel populations, including endangered and threatened species, in streams that flow into the mainstem rivers. Together, the databases and models can be used to make better decisions regarding how future growth and water use will affect water availability for diverse uses.

Colorado River Basin

The Colorado River is a critical source of water supply for much of the southwestern United States. The river supplies water to more than 25 million people, irrigates more than 3 million acres of cropland across seven “Basin States,” and supports aquatic and riparian ecosystems. While surface water in the Colorado River Basin is carefully regulated and monitored, other components of the water budget are less well understood. Through stakeholder consultation, the USGS has identified the following major components of the basin water budget for investigation: (1) estimation of current water use – in particular the consumptive use of water – and historical trends in water use into the future, (2) regional and field scale assessments of evapotranspiration and the dynamic variation in snowpack water content, and (3) estimation of groundwater discharge to streams and rivers. Better quantification of these components of the basin water budget will provide water managers with the increased knowledge of water sources and water movement and enhance their ability to make resource management decisions. For example, preliminary studies by the USGS indicate that somewhere between 20 and 60 percent of the surface- water flow in the upper basin is derived from groundwater discharge. Identification of stream reaches that receive large amounts of groundwater discharge will be a major effort in the upper portion of the Colorado River Basin during the Geographic Focus Area Study.

Preliminary USGS studies estimate that 20 to 60 percent of surface-flow in the upper Colorado River basin is derived from groundwater discharge.

Delaware River Basin

The Delaware River Basin covers parts of four States (New York, New Jersey, Pennsylvania, and Delaware). The basin has the largest inter-basin withdrawal of water east of the Mississippi River and provides water to over 15 million people. The information, databases, and products developed as a part of the Delaware River Basin Geographic Focus Area Study will contribute significantly to the information needs of the Delaware River Basin Commission "Strategy for a sustainable future water supply." Products will include (1) a database of water withdrawal, use, and return flow information for watersheds that will be accessible to water- resource managers in the basin; (2) a web-based tool developed using index streamgages to estimate daily streamflow from 1960 to 2010 for ungaged streams in the basin; (3) an evaluation of water needs for aquatic ecological systems within the basin including an updated decision support system for sections of the river, and development of flow and aquatic assemblage response relations for tributaries; and (4) a hydrologic model of the non-tidal portions of the watershed tributaries that can be used by water resource managers through an easy to use interface to evaluate potential impacts from future population, land-use, or water-demand scenarios.

> "The Census is a great example of leveraging resources for both the benefit of regional water management and the National Water Use and Availability Initiative. The Delaware River Basin Commission is getting invaluable assistance in the form of customized work products that advance our long-term Strategy for Sustainable Water Resources, and the USGS is getting very specific integrated water resource management feedback from a data rich basin that will inform and ground truth the National Water Census – to ensure that it is a value added to water managers across the Nation."
>
> —Robert Tudor, Deputy Executive Director Delaware River Basin Commission

Surface Water

The USGS operates over 7,000 streamgages to provide information on floods, droughts, and current water availability across the United States. This network of streamgages provides real time information and historical context for water resources planning and assessment. For studies of trends in streamflow and water availability, long historical records are critical. While extensive, the streamgage network cannot provide direct observations of streamflow at every location of interest. The Census aims to improve upon the information that is currently available at ungaged locations by providing estimates of daily streamflow at watersheds nationally through a point and click web application. A second major goal of the surface-water component of the Census is to evaluate trends in streamflow over time from a wide variety of causes, including changes to water-management strategies, land-use changes, and climate variability or change.

Groundwater

Groundwater presents a particular challenge in providing a complete budget on a watershed basis. Aquifer systems are complex, three- dimensional geologic features. They can cover great distances, commonly do not conform to surface-water divides, and may obtain most of their recharge at locations far from where the water discharges to a well or surface-water body. Because of these and other complexities, groundwater systems are incorporated into the Census in two ways. A major element of the Census focuses on regional analyses of groundwater availability in principal aquifers that collectively account for more than 90 percent of the Nation's total groundwater withdrawals. These studies are underway as part of the USGS Groundwater Resources Program. In addition, to the extent possible, estimates of groundwater recharge, storage, and discharge at the watershed scale will be made using a combination of information from the large-scale studies, data from observation well networks, analysis of streamflow records, and other available information.

Evapotranspiration

Evapotranspiration is an essential component of water budget determinations for water availability. It is also a fundamental variable of water use, especially for irrigation, with great importance for administration of water rights and river basin compacts. Historically, reliable quantification has required observations at sites having specialized instruments. However, these sites represent only local conditions, making evapotranspiration difficult to quantify over broad areas such as irrigation districts, river basins, or States. The USGS is applying its satellite remote sensing resources and expertise to quantify evapotranspiration across the landscape. The Geographic Focus Area Studies provide the context for testing evapotranspiration remote sensing methods for eventual application across the country to address both broad water availability and specific irrigation water use purposes.

There is an established evapotranspiration remote sensing community in the West with a history of estimating crop water use with Landsat imagery. Universities, consultants, State agencies, and Reclamation have all contributed to the development and application of techniques of this kind. To avoid duplication of effort and maximize contributions by various agencies to the Water Census, the USGS has entered into a partnership with Utah State University to develop specifications and guidelines for estimating crop evapotranspiration in the West with remote sensing. The vision is for a prescribed framework for inputs, techniques, and proven model performance, within which various entities could employ the model of their choosing, and publish crop water-use figures that would be recognized and accepted by the broad community of Western water stakeholders. In this way, important economies could be realized, while at the same time meeting standards for accuracy and precision for the Census.

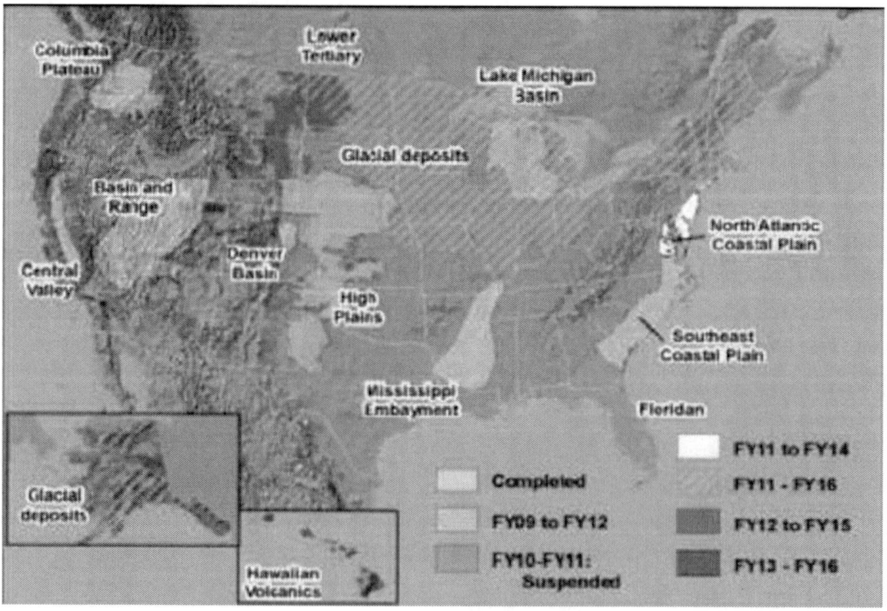

A map of USGS Groundwater Resources Program regional groundwater study areas, including the fiscal years during which the studies are planned to occur.

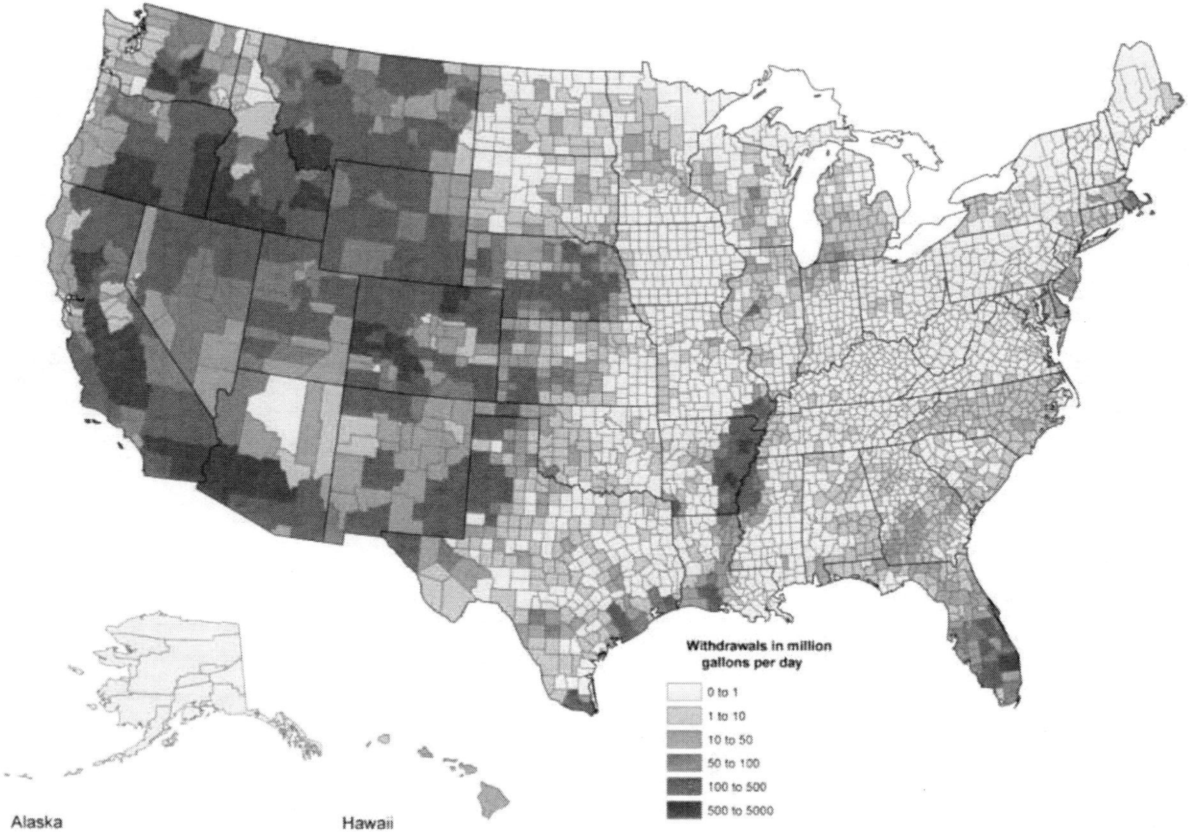

Freshwater irrigation withdrawals by county, 2005.

The Central Valley of California is another area where evapotranspiration remote sensing methods are being tested, leveraging ongoing USGS research on water productivity, and addressing the need for better quantification of fallowed land extent. In the first case, both yields and water use are being mapped, field by field, to reveal variations in the amount of "crop per drop," a valuable refinement to water-use reporting. In the second case, the objective is to better characterize the impacts of water shortages on land and water use by irrigated agriculture, a significant information gap identified by the California Department of Water Resources and other stakeholders participating in the National Integrated Drought Information System California Pilot, a NOAA-led activity.

Water Use

Information on human water withdrawals and return flows is critical information for the demand part of the water budget. Better information about components of water use such as withdrawal, conveyance, consumptive use, and return flow by sector of use and the factors which influence these components will allow water managers and users to make better decisions for the future.

Every five years since 1950, the USGS has produced a report on *Estimated Use of Water in the United States*. The most comprehensive of these reports was for 1995 and provided estimates of withdrawals, public-supply deliveries, wastewater returns, and consumptive use for 11 sectors of water use. The scope of subsequent reports did not include some sectors, and estimates of consumptive use. In addition, reports subsequent to 1995 estimated water use for counties but not watersheds, as previously done. The omitted components reduce the usefulness of this periodic assessment of the Nation's water use. As part of the Census, the USGS plans to return to the full assessment as represented by the 1995 report.

Efforts also are underway to improve estimation methods for the two largest sectors of the Nation's water use: thermoelectric power and irrigation. Thermoelectric power represented 49 percent of U.S. withdrawals in 2005, although significant amounts of the water withdrawn is returned to a surface-water body and is readily available for other uses. The USGS is developing improved estimation techniques for powerplants utilizing data reported to the U.S. Energy Information Administration. The two agencies are working together to improve the quality and usefulness of the reported data. Water withdrawals for irrigation made up 31 percent of 2005 withdrawals. Compared to thermoelectric power, a larger proportion of the water withdrawn for irrigation is consumed through evapotranspiration or incorporated into the crop, leaving less available for immediate reuse. The USGS has initiated several studies to analyze and improve estimation methods for irrigation water use.

A major goal is to refine the scale of water-use information. Efforts are underway to develop a National site-specific water-use database as part of the USGS National Water Information System, which will contain withdrawal, conveyance, use, and discharge information. Withdrawal locations, conveyances, and water system information is being entered for the public supply sector initially. Thermoelectric power plant locations and withdrawals will be added to the database, followed by other sectors for which site-specific information can be developed. Detailed, long- term information on withdrawals, conveyances, consumptive use, and return flows will provide the critical demand component for studies of the interactions between human water use and the natural hydrologic system.

Ecological Water Science

Ecological water science has advanced greatly over the past two decades driven largely by a need to better understand the relation between ecological water uses and the environmental flows required to maintain those uses, prevent degradation of freshwater ecosystems, and better maintain a balance between human and ecological water needs. The Census has a distinct role to fulfill in ecological water science and this includes developing novel tools and web-available resources that provide stakeholders and ecological water science practitioners with the hydrologic and biological information necessary for comparing natural and altered hydrologic regimes and understanding the effects of streamflow alteration on aquatic ecosystems. This approach is being applied at two distinct scales—nationally and at the large river basin scale as part of the Geographic Focus Area Studies.

In support of ecological water science at the national scale, the Census is developing and comparing flow modeling tools for building a national hydrologic foundation of baseline hydrographs that will ultimately provide hydrologic statistics for all ungaged streams in the United States. This information will be served directly to stakeholders via a map-based National Data Portal. This data portal will enable connections to other large data compilations including those collected by other Federal, State and Tribal organizations.

In addition to the national scale efforts, ecological water science is part of each of the Geographic Focus Area Studies at the large river basin scale. Ecological water science activities in the Apalachicola- Chattahoochee-Flint combine basin-wide streamflow models with on- the-ground measurements of changes in the occurrence or abundance of different kinds of fish and mussel species. In the Colorado River Basin, the Upper Verde River is serving as the first sub-basin for ecological water analysis. The analysis includes determining the hydrologic record, how far the seasonal hydrology departs from the historic record, the amount of habitat needed to support different types of aquatic species under different flow volumes, a status of the native and nonnative species, and a water budget including groundwater that helps identify reaches vulnerable to drying. Competing water needs in the Delaware River Basin, the longest undammed river east of the Mississippi, make it ideal for developing tools for sustainable management and water conflict resolution. Field and laboratory experiments are underway to develop habitat suitability criteria that model the relation between physiological response of key species and changes in temperature and hydrology.

Fish assemblage sampling in the upper Apalachicola-Chattahoochee-Flint River Basin. Field data will inform models that can be used to simulate biological responses to changes in water availability.

Research findings and modeling tools developed as part of the Geographic Focus Area Studies, as well as the national-scale effort, will be useful to State and Federal natural resource agencies and other stakeholders charged with ensuring that water management actions are consistent with meeting human needs and protecting biological integrity as mandated by the Clean Water Act and with conserving imperiled and (or) recreationally valued species. Developing modeling tools with a high degree of transferability is an important element of the Census and the research, tools, and applications being developed as part of the ecological water science efforts are expected to have applicability well beyond the borders of the study areas in which they were developed.

Planning for the Future

Progress toward a Census has been made by integrating information from a number of programs. Over the next few years, progress will continue with existing efforts and expand as funding allows. In addition to completing the work described above, priorities for future work (not in priority order) are as follows:

Expand Geographic Focus Areas–Geographic Focus Area Studies provide an opportunity to prototype approaches and leverage and support ongoing water availability studies. The three current Geographic Focus Area Studies will be completed in 2014. Future Geographic Focus Areas will be identified using criteria being developed as part of the implementation plan for the USGS strategic science direction on water resources.

Evaluate impaired surface water and groundwater that may be important future sources of water supply–The National Water Availability and Use Assessment includes an assessment of "impaired surface water and groundwater supplies that are known, accessible, and used to meet ongoing water demands." Likewise, the SECURE Water Act calls for a brackish groundwater assessment. Since passage of the Act, the USGS Groundwater Resources Program undertook preliminary work developing approaches to brackish water assessment in three areas.

Continue to enhance ways to bring information together from multiple sources–It is critical that the Census use information available from multiple sources to the extent possible. The Census Program will continue to work through the Federal Advisory Committee on Water Information to enhance this capability, including possibilities for data portals that provide ready access to data from multiple agencies, such as proposed by the advisory group's Subcommittee on Ground Water.

Work with State and other agencies to enhance water-use datasets– Studies summarizing water use highlight limitations and inconsistencies in water- use data collection and reporting across the United States (National Research Council, 2002; U.S. Government Accountability Office, 2009). Reported water use relies heavily on estimates by State agencies, and estimation procedures vary from State to State. The Census provides an opportunity to establish agreed upon standards for estimating, reporting, and storing water use data from various categories. In addition, USGS has accentuated efforts to work with other Federal agencies such as the Department of Agriculture and the Department of Energy, particularly with respect to water-use data related to thermoelectric power generation and irrigation water use.

Identify areas where there is the most uncertainty in various types of data– The Census must focus efforts on the most critical areas rather than spread its effort equally over the landscape. One factor to consider in focusing efforts is where the uncertainty in different types of data is greatest. The Census will use uncertainty analysis from models to help address this issue.

SECURING A SUSTAINABLE WATER FUTURE

WaterSMART provides the framework Interior needs to successfully join the related water resource programs of its bureaus and offices in pursuit of a single unifying purpose: to secure a sustainable water future for the nation. Solutions to complex water resource challenges now and in the future will require finding balance among the many uses of water that support people, the economy, and the environment, and between the supply of water and the demand for water. Likewise, a balanced strategy for sustainable management of water resources includes efforts to conserve existing supplies through efficiency and reduced demand, developing available sources of water, planning for an uncertain future in the face of a changing climate, and integrating water and energy policies. By building on the success of

WaterSMART, Interior will continue providing leadership and assistance to partner states, tribes, regional and local governments, and conservation organizations as a responsible steward of the Nation's natural resources.

For more information, visit www.usbr.gov/watersmart.

INDEX

C

F

G

H

N

O

P

Q

R

S

T

U

V

W

Y